BOTANICAL RESEARCH AND PRACTICES

NEW DEVELOPMENTS IN ALLELOPATHY RESEARCH

BOTANICAL RESEARCH AND PRACTICES

Additional books in this series can be found on Nova's website
under the Series tab.

Additional e-books in this series can be found on Nova's website
under the e-book tab.

BOTANICAL RESEARCH AND PRACTICES

NEW DEVELOPMENTS IN ALLELOPATHY RESEARC

JULIA E. PRICE
EDITOR

nova publishers
New York

NOTICE TO THE READER

Additional color graphics may be available in the e-book version of this book.

Library of Congress Cataloging-in-Publication Data
New developments in allelopathy research / editor: Julia E. Price.
pages cm. -- (Botanical research and practices)
Includes index.
ISBN 978-1-63483-390-5 (hardcover)
1. Allelopathy. 2. Allelopathic agents. I. Price, Julia E., editor. II. Series: Botanical research and practices.
SB732.75.N49 2015
632'.96--dc23
2015025935

Published by Nova Science Publishers, Inc. † New York

CONTENTS

PREFACE

Allelopathy refers to the beneficial or harmful effects of one plant on another, including both crop and weed species. Mushrooms is one such example which provides various effects to their curcumjacent plants by releasing allelochemicals. The authors' of this book examine the allelopathy of mushrooms and showed that all mushrooms have allelopathic activity without exception. Furthermore, rice, barley, wheat, rye, sorghum and sunflower are known to be allelopathic crops. The authors focus on buckwheat in particular, to determine whether or not it is allelopathic as well. Other chapters examine the allelopathic effects of macrophytes in subtropical shallow lakes; and the invasive potential of parthenium weed, a herbaceous plant, and one that has now invaded over 44 countries worldwide, having an impact upon both crop and pasture production, natural community biodiversity and human and animal health.

Chapter 1 – Mushrooms, which are also called as higher fungi, give various effects to their circumjacent plants by releasing allelochemicals as same as usual allelopathic plants. However, reports focused on the mushroom allelopathy are rarely to be found, and the studies indeed have been not developed enough up till now. In this chapter, previous reports concerning mushroom allelopathy will be reviewed, and recent results revealed by the author will be introduced in addition. In the plant growth inhibition assay of over 500 fruiting bodies to lettuce (*Lactuca sativa* L. var. Great Lakes 366) seedling growth, all mushrooms showed allelopathic activity to a greater or lesser degree. The result showed that all mushrooms have allelopathic activity without exception. On the other hand, differences of activity were observed among same species. This findings indicated that there is a possibility a kind of allelochemicals and the quantity of them are variable in some species. Further, a daedal mechanism of allelopathy of *Cryptoporus volvatus*, which is resemble to the allelopathy of black walnut (*Juglans nigra*), will be introduced in detail.

Chapter 2 – In a growth chamber trial the suppressive potential of common buckwheat (*Fagopyrum esculentum* Moench) on redroot pigweed (*Amaranthus retroflexus* L.) was tested by separating resource competition for light, water and nutrients from belowground root interactions between the two species. The growth repressing effect of buckwheat on redroot pigweed was very strong (>89%) and due to both competitive shading effects and root interactions of potentially allelopathic nature. In order to verify this result in situ a field trial was performed in 2014. Buckwheat and redroot pigweed were sown together on the 7th of August and the dry weight of redroot pigweed was determined 28 and 55 days after sowing (DAS). A strong repression of redroot pigweed growth by buckwheat could be observed at both dates independently of shading. When soil samples from the field within the buckwheat stands 11, 26 and 42 DAS were transferred into the growth chamber no effect on redroot pigweed and lettuce (*Lactuca sativa*) growth could be observed. Assuming that allelopathic molecules must be present in the soil solution supplementary experiments were conducted in the phytotron. Under controlled conditions buckwheat was grown in pots for different periods of time. These pots were used for several growth experiments with lettuce. Lettuce root length was measured after exposing seeds to fresh, dry soil from the pots, soil solution samples collected with Rhizon soil moisture samplers and boiling water extracts of the soil. Moreover, buckwheat and lettuce were sown at the same time in petri dishes in soil and on blotting paper; subsequently lettuce root length was measured after 3 days. In none of the experiments an influence on lettuce and redroot pigweed development could be observed. In addition, redroot pigweed and buckwheat roots were *separated with a permeable membrane in a pot trial allowing molecules to diffuse between the compartments for 28 d. Redroot pigweed grown in the membrane was of the same size than in the control without buckwheat.* The authors therefore conclude that there are either no allelopathic molecules in the soil solution (not soluble in water) or that they are rapidly degraded, roots must be in direct interaction to allow allelopathic effects, the growth inhibiting effect is due to a long term and constant exposure of small quantities of allelopathic molecules or the growth repression of redroot pigweed by buckwheat is not due to allelopathy.

Chapter 3 – Autotoxicity is a common problem in strawberry. During this phenomenon strawberry plant's roots secreted allelochemicals to the culture solution causing damage to the root cells, which in turns hamper water and mineral nutrient absorption. As a result, the growth of shoot and root, number of flowers and harvested fruit per plant, and fruit enlargement reduced greatly. In their two studies the authors have tried to recover the growth and yield of strawberry by the foliar supplementation of amino acids and electro-degradation (ED) of recycle nutrient solution. In both studies, the authors found the recovery of growth and yield of strawberry during autotoxicity. In our first study, twenty two amino acids were short listed to Ala, Glu, Hyp, Thr, His and Phe on basis of their performance on growth and yield of strawberry in non-recycled hydroponics. Then short listed amino acids and GABA were further investigated following Wagner's pot hydroponics. Among the seven amino acids applied, only Glu and Hyp improved their fruit yield over 50%. The *in vitro* culture of strawberry plantlets showed similar results of previous experiments. Therefore, considering the effects of amino acids on growth and fruit yield of strawberry, Glu and Hyp can be used for foliar application in a non-recycled hydroponic culture. In our second study, the authors found complete recovery in fruit yield of strawberry plants grown in closed hydroponic system where culture solutions were allowed to ED treatment before reuse. In the growth chamber bioassay strawberry plantlets showed retarded growth in non-renewed and/or its weekly ED whereas it was recovered in biweekly applications. Therefore, ED nutrient solutions were investigated following without plant experiments. It was found that ED of nutrient solution was associated with degradation of Fe-EDTA (about 10%) and low concentration of Ca^{2+} (about 66% of the initial) in the treated culture solution. The disadvantages of longer duration ED were decrease in solution pH as low as 3.1 and increase in temperature. Therefore, shorter duration (2h) ED treatments were applied in the greenhouse experiments following Wagner's pot hydroponics. ED machine used for previous experiments was modified for allowing the treated solution passed continuously through the electrode. Results obtained from Wagner's pot experiment showed that growth and yield (about 99% to renewed control) of strawberry plants were improved in non-renewed nutrient solution having

ED at four weeks interval. Therefore, in practice, application of ED to non-renewed nutrient solution in closed hydroponic system for 2 h at every four weeks intervals would recover autotoxicity in strawberry.

Chapter 4 – The globally successful invader *C. canadensis* is the first species to colonise the sandy levees beside the rivers into new areas and, owing to high levels of seed production and in particular its allelopathic effect, it completely dominates the other species in the ruderal phytocoenosis. The content of total phenolics and phenolic acids varies, following the order: vegetative plant parts > dead plant parts > sandy soil under *C. canadensis*. Water leachate and soils inhibited seed germination and seedling growth of the neighbouring herbaceous plantsto varying degrees: vegetative parts > dead parts > sandy soil, which is directly related to the content of total phenolics and phenolic acids in them. Secondary metabolites, above all phenolic compounds released from its vegetative organs, reach the sand under it, where they have a negative effect on seed germination and seedling growth of neighbouring herbaceous plants, reducing their abundance and cover. The dynamics of allelopathically significant phenolic compounds in *C. canadensis* L. tissues and in the sandy soil have a marked seasonality and coincide with certain phases in plant development. The highest amount of phenolics in the tissue was measured during the phase of plant elongation and intensive growth, and again during the phase of fruit abscission and plant decline. It can be seen that the level of free acids in the soil under *C. canadensis*, increases dramatically between March, when the plants are in the form of rosettes, 1-2 cm in diameter, and July, the beginning of flowering. In these months free phenolic acids account for as much as 73.7% of the total free phenolics in the soil. The percentage of soil bound phenolic acids gradually increases between April, when the plants are in the form of larger rosettes, 5-10 cm in diameter, and August (in full flower and beginning of fruiting), when their maximum level (43.5%) is reached. Phenolic acids can serve as reliable bioindicators of phenolics metabolism and turnover, starting from their biosynthesis in plant tissues to their incorporation into stable fractions of soil humus. Also, they show the level of improvement in substrate quality for further plant colonisation in comparison to sands. An important role in the spread of this invasive species into new areas and its domination over native species is played by phenolic compounds as the most abundant group of allelochemicals. These exert regulatory roles in many processes, not only in plants themselves, but as well in the ecosystem as a whole.

Chapter 5 – In subtropical shallow lakes, due to shallowness and mild climate, macrophytes grow continuously and colonize extensive areas. Many macrophytes are known for the production and release of allelochemicals, with effects extensively reported for phytoplankton and cyanobacteria, but poorly elucidated for heterotrophic bacteria. In this chapter, the authors reunited three studies conducted in subtropical shallow lakes in southern Brazil, where many allelopathic macrophytes occur, such as *Potamogeton illinoensis, Potamogeton pectinatus, Cabomba caroliniana, Egeria densa, Myriophyllum spicatum* and *Ceratophyllum demersum*. In the first study, the phytoplankton community and environmental variables were assessed in 5 seasonal campaigns through transects ranging from a macrophyte-vegetated to a macrophyte-free area in the large Lake Mangueira. The chlorophyll *a*, cyanobacterial biomass, the total biomass, diversity and richness of the phytoplankton community, as well as several nutrients were lower in the macrophytes dominated part of the transect and increased in the macrophytes-free area. In the second study, microcosm coexistence experiments with *C. caroliniana, E. densa, M. spicatum*, and *C. demersum*) and the cyanobacteria *Microcystis aeruginosa* under different levels of phosphorus concentrations have revealed that, although all plants were effective competitors for nutrients and impacted negatively the *M. aeruginosa* biomass, only *M. spicatum* and *C. caroliniana* provided evidence of allelopathic inhibition. In the third study, the allelopathic effects of *P. pectinatus* on the microbial food web (heterotrophic bacteria, flagellates and ciliates) and phosphatase activity were measured in the field in areas with and without the macrophytes and also in mesocosm experiments with the addition of methanolic extracts, decomposing plant parts and water surrounding the stands. Field results showed lower microbial food web components densities as well as enzymatic activity in areas with the presence of the macrophytes, whereas in the experiments the methanolic extract and the decomposing stems and leaves induced a one day longer lag phase in the bacterial growth and lower rates of enzymatic activity when compared with the control. Taken together, these results provide further field evidence for the allelopathic effects of macrophytes on phytoplankton, autotrophic and heterotrophic bacteria, corroborating much of the literature available so far in this topic, as well as provide a mechanistic explanation for the lower bacterial density, biovolume, biomass, production and respiration often found in areas where these macrophytes occur. Allelopathy adds, hence, another layer of complexity to the already known strong structuring role of macrophytes in shallow lakes, imparting poorer functioning of microbial food webs in areas densely colonized by these plants.

Chapter 6 – Parthenium weed (*Parthenium hysterophorus* L.) is an herbaceous plant that has now invaded over 44 countries worldwide having an impact upon crop and pasture production, natural community biodiversity and human and animal health. Allelopathy is a well-defined physiological trait that is possessed by some crops and native plant species, but especially by invasive weeds. It is a biological phenomenon involving the release of chemicals that may cause a stimulatory, but more often an inhibitory effect upon another plant's growth, reproduction and survival. Numerous studies have associated the invasive success of parthenium weed with a proficient allelopathic attribute, able to cause an inhibition of either germination, seedling or plant growth, phenological development and/or flowering in a range of agriculturally important plant species, including weeds and some native species. The presumed allelochemicals are produced by a number of plant parts including the stem, leaves, roots, pollen and other floral parts. The two major allelochemical groups that are thought to be involved are the sesquiterpene lactones and the phenolic acids, and both have been found to be released by living plants and decomposing plant residues as well as being present in rhizosphere soil under a parthenium weed population. Different bioassay approaches have been used to help illucidate the allelopathic nature of this weed. These include laboratory assays using aqueous extracts applied directly to germinating seeds, or those evaluating the impact of parthenium weed seedlings on the growth of neighbouring seedlings, both using a liquid medium (relay seeding bioassay) or in an agar medium (plant box bioassay). Other laboratory bioassays have looked at the impact of parthenium weed litter upon germination and seedling growth (sandwich bioassay) while some glasshouse pot tests carrying sand or soil have also been undertaken looking at the impact of aqueous extracts, leaf litter or rhizosphere soil. Each approach has its own advantages and disadvantages. This chapter will evaluate the evidence and decide if allelopathy makes a major or a minor contribution to the invasiveness of this weed.

In: New Developments in Allelopathy Research
Editor: Julia E. Price

ISBN: 978-1-63483-390-5

Chapter 1

ALLELOPATHY OF MUSHROOMS

Hiroshi Araya*
Graduate School of Agriculture, Meiji University,
Tama-ku, Kawasaki, Japan

ABSTRACT

Mushrooms, which are also called as higher fungi, give various effects to their circumjacent plants by releasing allelochemicals as same as usual allelopathic plants. However, reports focused on the mushroom allelopathy are rarely to be found, and the studies indeed have been not developed enough up till now. In this chapter, previous reports concerning mushroom allelopathy will be reviewed, and recent results revealed by the author will be introduced in addition. In the plant growth inhibition assay of over 500 fruiting bodies to lettuce (*Lactuca sativa* L. var. Great Lakes 366) seedling growth, all mushrooms showed allelopathic activity to a greater or lesser degree. The result showed that all mushrooms have allelopathic activity without exception. On the other hand, differences of activity were observed among same species. This findings indicated that there is a possibility a kind of allelochemicals and the quantity of them are variable in some species. Further, a daedal mechanism of allelopathy of *Cryptoporus volvatus*, which is resemble to the allelopathy of black walnut (*Juglans nigra*), will be introduced in detail.

INTRODUCTION

Fresh wild mushrooms are recognized as autumn foodstuffs. However, we can obtain some kinds of them as dried, freeze, oiled or pickled states irrespective of seasons. Further, since cultivation methods for several mushrooms have been established, we can purchase them all the year around at markets. The season of wild mushrooms occurrence is generally limited, and the period is not so long. In addition, the amount of mushrooms occurrence depends on various factors, such as climate, location, nutrients, etc.

---

* Corresponding author: Tel/Fax: +81-44-934-7832, E-mail: kinoko@meiji.ac.jp (H. Araya).

Mushrooms are classified as fungus produce fruiting bodies could be observed visually, and it is different from the biological classification. The mushrooms are composed from Basidiomycetes mainly, and Ascomycetes secondary. To date, twenty thousand over mushroom species have been reported in the literatures, and the actual number of species of mushrooms has been estimated as from one hundred thousand to two hundred thousand, which correspond to ten percent of total fungal species [1]. Even the size of fruiting bodies are visible to the naked eye and those are able to be collected relatively easily, listed species are much fewer than substantial species like as other class of microbes. The fact, that the existence periods of the fruiting bodies of majority of mushroom species are short time (from a few days to fortnight, typically), might make the classification of mushroom species difficult. The construction of inventory of mushrooms combined with specimens and the information of their genome has been driven presently, there is a possibility to major change the estimation of the number of mushroom species.

It has been well known in the field of natural products chemistry that mushrooms produce various unique secondary metabolites, which are useful source for searching lead compounds to develop new chemicals, such as pharmaceuticals, agrochemicals, food additives, and so on [2]. Former researchers had isolated various interesting chemicals, and elucidated their structures. However, the number of reports about wild mushroom fruiting bodies is fewer than that of plants in fact, because sufficient amount obtention of fruiting bodies is generally. On the other hand, some species including inedible mushrooms have been cultivated as mycelium state (and rarely as fruiting bodies), and their metabolites have been investigated. However, it must be noted that there is differences in metabolites between fruiting bodies and fermented mycelium in many cases. The fact indicates that chemical ecological studies in nature need to handle naturally occurring fruiting bodies.

The detail life cycle of mushrooms is withdrawn in this chapter. As all know, mushrooms are made with mycelium or those aggregate except the peculiar cells such as a spore or the basidium and if the density of mycelium of the diploid becomes higher, fruiting body rudiments form and grow up to the fruiting bodies. It is thought that an allelopathy is easy to be observed the place where fruiting bodies exist if there are few weeds at a place where a certain mushroom occurs for short period. This is a big advantage in the study of mushroom allelopathy about microorganisms. However, because the fruiting bodies occurring period of the wild mushrooms is usually short period except some kinds such as Polyporaceae fungus, it is necessary to happen to meet it in the timing when, fortunately, fruiting bodies occur even if the mushroom forms a big demesne. The reason why the mushroom allelopathy study does not advance in comparison with other plants allelopathy study is the difficulty in timing for such an observation. Fortunately, the wild mushroom hunting, which is one of hobbies of the author, has lead findings of mushroom allelopathy in wild state. It is the opportunity when this would be engaged for mushroom allelopathy research.

The review in this chapter is rewritten recent papers in Japanese [3].

History of Mushroom Allelopathy Study

There are reports concerning plant growth regulating substances produced by mushrooms. However, there are only five reports focused on mushroom allelopathy by four groups except for the author's works [4-10]. This may be usual appearance, because even study on plant allelopathy is lesser than all of phytochemical study.

First mushroom allelopathy was reported by Hans Molisch, who was the pioneer of allelopathy in 1937 [11]. He investigated an influence of *Agaricus campestris* fruiting bodies to growth of *Vicia sativa* L. in a sealed bottle, and had a result that the mushroom inhibited the growth of the plant slightly. The cause depends on volatiles emitted by the mushroom apparently. To date, some groups reported that ethylene, a plant hormone, has been produced by several mushrooms and undermentioned 1-octen-3-ol has been detected from many mushrooms. These compounds could act as allelochemicals in the experiment.

Chaumont and Simeray reported the influence that the water extract of 114 kinds of mushrooms gives to the germination and growth of the red radish *Raphanus sativus* var. radiciola in 1982 [12]. By performing the assay with extracts, which prepared using four times distilled water to fruiting bodies weight, they obtained a result that 1) there were 20 kinds of species indicating complete germination inhibition, 15 kinds of species inhibited more than 80%, and 48 kinds species inhibited 50-80% of the germination, 2) most of the extracts inhibited the red radish seedling growth remarkably.

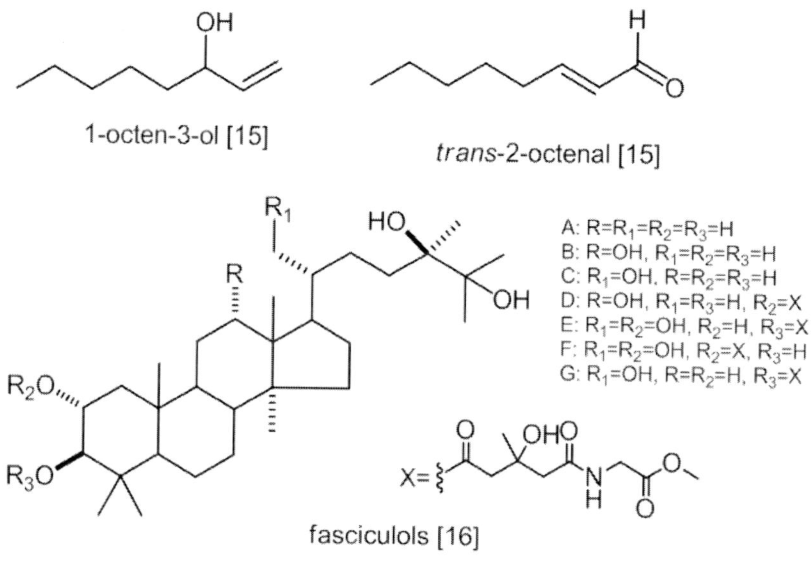

Figure 1. Compounds considered as allelochemicals because of observation of weed suppression.

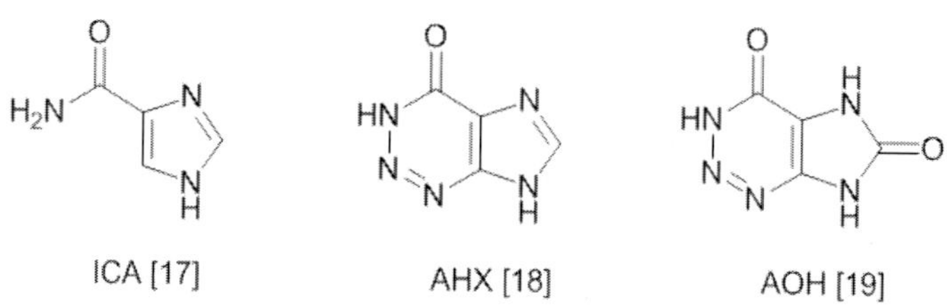

Figure 2. The Chemical Structures of Fairy Chemicals [20] from *Lepista sordida*.

A similar studies of that were reported by the Uzun, et al., [13] and Mo, et al. [14], too. Uzun, et al. reported that methanol extract of ten kinds of mushrooms inhibited corn (*Zea mays*) seeds germination and sprouting seeds growth in 2010. Since the examined extracts were prepared at high density relatively, all significantly showed strong activity. Mo, et al. reported that ethanol and water extracts of *Lactarius hatsudake* greatly inhibited the growth of the sprouting of *Oryza sativa* and *Echinochloa colona*. According to an observation of the author, most of the place found *L. hatsudake* fruiting bodies is relatively young red pine (*Pinus densiflora*) groves mixed together with Japanese pampas grass (*Miscanthus sinensis* Andersson) belongs to the same Gramineae plant in Japan. As there is not influence for the growth of the Japanese pampas grass of the fruit body scene neighborhood obvious allelopathy of the fungus is not appeared, it thought that there would be almost nothing allelopathy to Gramineae in the wild Japanese field.

Allelochemicals as gas state were only determined in black Truffle (*Tuber melanocporum* Vitt.) by Streiblová, et al. in 2012, which is called as one of the three major delicacies of the world [15]. The mycorrhiza mushroom can be produced by semi-artificial cultivation which is the transplantation of young willow oak trees infected and formed mycorrhizae by the fungus. However, it takes more than ten years until first harvest of the fruiting bodies, and no one can predict the time. Interestingly, coming to produce of the fruiting bodies in the underground can be judged easily, because weeds growth are suppressed remarkably to the above ground part of the fruiting bodies. As a matter of course, the fragrance components of the mushroom have been analyzed by several researchers, and more than 200 kinds of volatiles have been identified. But they determined 1-octen-3-ol and *trans*-2-octenal (Figure 1) as volatiles having strong plant growth inhibition activity. The allelochemical 1-octen-3-ol which is well known as a major fragrance of *Tricholoma matsutake* was identified, and is one of the compounds which most mushrooms produce. It is thought that mycelium growing up in the underground produce these compounds, and released these compounds near the rhizosphere act to plant roots effectively as growth inhibitor, because the diffusion is extremely small. In addition, since it is confirmed that the plant hormone such as ethylene or indole acetic acid have been produced, the black truffle might show allelopathy by releasing several allelochemicals.

In addition, Ikeda et al. reported fasciculols (Figure 1), which were isolated as growth inhibitors from fruiting bodies of *Naematoloma fasciculare* [16]. They did not use the word 'allelopathy' in the paper. But, since they start the work for determination of plant growth inhibitors on the basis of the authors observation that there was few weeds around of the fruiting bodies, fasciculols could be agreed as allelochemicals.

Extremely in late years, Kawagishi et al. clarified the substances which affected a phenomenon 'Fairy ring' to participate in allelopathy. They isolated two kinds of plant growth regulators from culture broth of *Lepista sordida* (Schum.: Fr.) Sing., which is a kind of so-called 'Fairy Ring Mushrooms' [17-19]. The fairy ring is a phenomenon that some mushroom fruiting bodies occurs into a ring form on the ground, and is handed down fairies danced in the ring literally in old Europe. This fairy ring occurs in the grassland such as golf courses and is known as fairy ring disease. The fairy ring on the turf which caught a disease continues extending the ring with age, but a color is dark, and the turf grows thick in the immediate outside of the ring which a fruit body produces, and the turf withers and dies in the immediate inside of the ring. To date the drug which is a special effect is not developed. Kawagishi, et al. cultured a discrete mycelium from L. *sordida* fruit body and identified imidazole-4-carboxyamide (ICA) [17] as a growth inhibitory compound and 2-

azahypoxanthine (AHX) [18] as a plant growth promoter (Figure 2). Furthermore, it was suggested AHX absorbed to a plant is converted into 2-aza-8-oxohypoxanthine (AOH, Figure 2) in recent report by a xanthine oxidase of enzyme in plants, and which is act as a plant growth promoter [19]. In addition, the presence of these AHX, AOH in usual plants has been confirmed by analysis by LC-MS/MS. This result showed that these compounds could be act as a plant hormone. The applied study for farm products increases in production is started and a result was reported that rice yield increases more than 20% by the dosage of these compounds. Recently, these compounds was named as 'Fairy Chemicals' after fairy ring phenomenon [20]. These fairy chemicals could be conceded as allelochemicals.

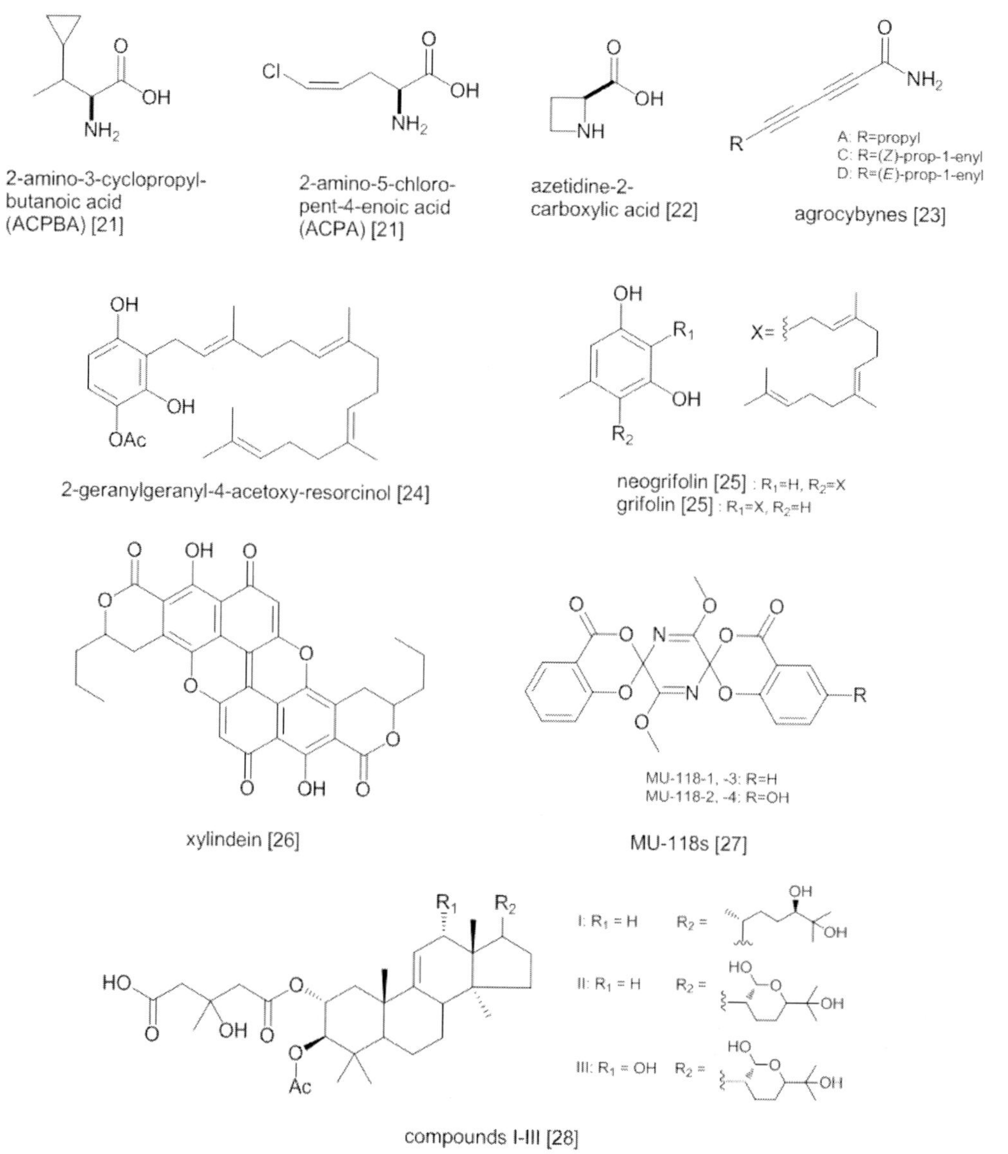

Figure 3. Chemical structures having plant growth regulating activity (allelochemical candidates) isolated from various mushrooms.

The all above reports were paid its attention to the mushroom allelopathy. To date except for the authors study, unfortunately, allelochemicals of mushrooms only determined in the example of the black truffle.

On the other hand, constitutional studies in mushrooms have been carried out all over the world, and dozens of kinds of interesting plant growth regulating substances have been isolated [21-28]. The chemical structures of them are showed in Figure 3. These compounds act as plant growth regulators in laboratory environment, and considered as allelochemicals. But there is no observation of the mushroom allelopathy in nature, so these compounds should be allelochemical candidates in nature.

In addition, there are other candidates of allelochemical produced by mushrooms. A number of secondary metabolites of mushrooms have been reported and most of them have been not described about its activity to plant growth. In some case, same chemicals were isolated from other organisms, and their plant growth regulatory activity was reported. For example, L-DOPA, which was deduced as an allelochemical of one of legume (*Mucuna pruriens*) [29], was detected in several mushrooms. Especially, *Strobilomyces floccopus* and *Hygrocybe conica* contained L-DOPA at highly-concentration, and these mushrooms were more likely to use L-DOPA as an allelochemical [30, 31]. Unfortunately, examples of the same case have been not investigated in enormous literatures.

The all above is information about mushroom allelopathy the author has up to now. The concrete knowledge about mushroom allelochemicals has been not provided, and further research and literature retrieval need for clarification of mushroom allelopathy.

MUSHROOMS HAVING STRONG ALLELOPATHIC POTENTIAL

A definition of the allelopathy does not have the description about the strength and weakness of the effect. In addition, as the experiment that Molisch performed was so, it is not limited to a phenomenon in the outdoors, and many researchers nowadays include the *in vitro* phenomenon of the laboratory scale as allelopathy too. All creatures produce secondary metabolite unlike the primary metabolite. Structures of the secondary metabolites are various, and all creatures revealed some effects more or less in their metabolism by receiving the compounds. In other words, it should be said that all creatures have allelopathy at different quality and degree, then, all mushrooms have allelopathy!

The studies conducted by above-mentioned Chaumont and Simeray [12], Uzun, et al. [13] and Mo, et al. [14] carried out bioassays using solvent extracts of fruit bodies. As for the solvent extract, the constituents could change by the polarity of solvents. In other words, a certain solvent extract may not contain some important allelochemicals in the original mushroom. As a matter of course, polar compounds should show activity, because water, such as rain, act as solvent in nature. However, it could not deny that less polar compounds play an important role in the allelopathy. In addition, like an example of juglon which is an allelochemical of black walnut (*Juglans nigra*), we must not forget that there is an example that a precursor of most quantity receives degradation by microbes and oxidation with air, and changes to the high potential allelochemical.

To evaluate potential of whole fruiting bodies of mushrooms, not samples of extracts, the author of this chapter applied "the sandwich method" [29] that is bioassay using the sample itself. More than 500 mushrooms tested their allelopathic activity to lettuce (*Lactuca sativa* cv. Great Lakes 366) seedlings. As a result, the following conclusion was obtained [10].

1) All mushrooms showed allelopathic activity.
2) Most of mushrooms inhibited radicles and hypocotyl growth of lettuce seedlings.
3) The allelopathic activity to the root showed normal distribution regardless.
4) In the comparison among genus, Laccaria, Ganoderma, the Hypholoma exhibited strong growth inhibitory.
5) *Lyophyllum decastes*, *Hygrophorus pudorinus* (Fr.) Fr., *Xeromphalina tenuipes*, *Cryptoporus volvatus* showed particularly strong growth inhibitory activity, and subsequently *Hebeloma spoliatum*, *Calocybe gambosa*, *Russula nigricans* (Bull.) Fr., *Lentinus lepideus* (Fr.:Fr.) Fr., *Suillus granulatus* (L.:Fr.) Roussel, *Entoloma sarcopum*, etc done.
6) Several species, *Amanita hemibapha*, *R. nigricans* (Bull.) Fr., *Boletus auripes* Peck., etc., showed varied allelopathic activity within same species.
7) *Suillus luteus* (L.:Fr.) S.F.Gray, *Exidia glandulosa*, *Inonotus obliquus*, *Fomes fomentarius* (L.: Fr.) Kickx, *Pleurotus ostreatus*, *Ischnoderma resinosum* (Fr.) Krast., etc., promoted either radicle and hypocotyl growth.

The above-mentioned result suggested that quality, and/or quantity of allelochemicals produced by mushrooms could be changed by many factors such as locality, climate, succession, etc. About the allelopathy of the mushroom to grow under environment of nature, it is necessary to confirm that desired mycelia of the mushrooms exist in the investigation area. The simplest method is watching of the occurrence of the fruit body, but it is difficult in many cases because those occurrence time and period are short and limited. Because easy determination of allelopathy from mushrooms to plants by the only viewing temporarily, it is a custom to observe the mushroom allelopathy with caution when the author strolled in forests. In infertile area not found living things such as plants with naked eyes, an allelopathy may be produced by a certain microorganism. In this case, isolation and identification of the causal microorganism, or metagenomics approach are indispensable. There is still far to go before identification of the allelochemicals. It is thought that much knowledge is provided by the development of the future study at all events.

ALLELOPATHY OF *CRYPTOPORUS VOLVATUS* [7, 32]

Cryptoporus volvatus is white rot fungus belong to polyporaceae, and the mushroom is distributed over Southeast Asia and the North America. The mushroom has been considered to be one species in one genus for long time, but recently *C. sinensis* occurring in china was recorded as one of new species in same genus. gathered in China is listed as a new kind [33]. Furthermore, *C. volvatus* growing in the North America is differ from other Cryptoporus species in the state of the fruit body as far as the author watches it with a photograph on the web, and thinks it is more likely to be another species. The fungus make fruiting bodies on a trunk of a few dying Pinaceae species. It is withering and dying or these mushroom wither and die and open a fruit body in the interval for lying scattered on a trunk of few conifer particularly Pinaceae. The fruiting body is non-poisonous, but unsuitable to eat because of their toughness and strong dried fish smell. There were extremely few weeds around the Japanese red pine which the fruiting bodies occur in large quantities in comparison with the outskirts of the same tree which the fruiting bodies do not occur. Because it was suggested that participation of the allelopathy due to the *C. volvatus* was different from pine tree allelopathy, [34] the author started the research on the mushroom allelopathy.

It's well known that the mushroom produces interesting secondary metabolites, cryptoporic acids (CPAs) which are composed from isocitric acid and albicanol through ether bond(s). As shown in Figure 4, CPAs are classified into three types based on their structural characteristics: the monomeric type - an isocitric acid and an albocanol are linked via an ether bond; the linear dimer type – two monomeric units are connected with an ester linkage; the cyclic dimer types - two monomeric units are connected symmetrically with two ester bonds. The secondary metabolites which isocitric acid included are rare compounds, and have been found in only CPAs analogue to date. It was expected that these CPAs would be isolated as allelochemicals by the bioassay guided fractionation, because Asakawa, et al. reported a series of CPAs inhibited germination and growth of the rice (*O. sativa*) at 200 ppm evenly [35].

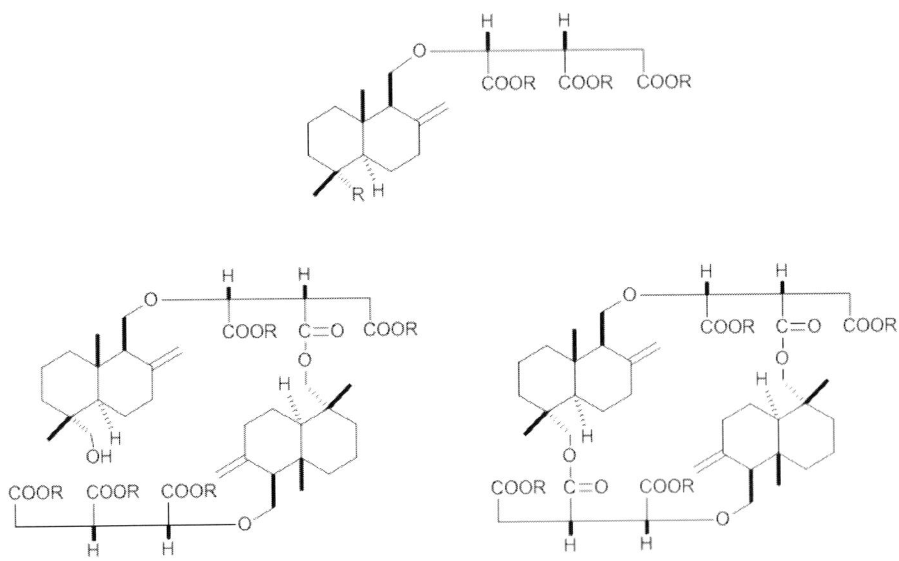

Figure 4. Three basic structures of cryptoporic acids (CPAs).
(R groups in the chemical structures means $-H$, $-CH_3$, $-CH_2CH_3$, $-OH$, or $-OAc$).

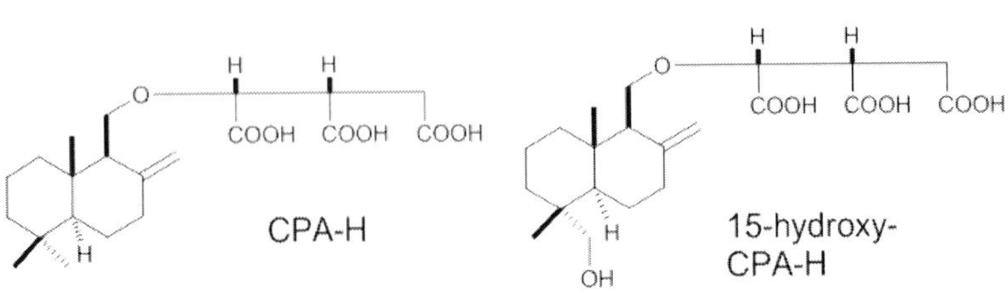

Figure 5. The chemical structures of two major allelochemicals in *C. volvatus* fruiting bodies.

Isolation strategy of allelochemicals was applied a concept of total activity [36] to isolate certain allelochemicals to get the compound which contributed to activity in the original material. Lettuce (*L. sativa* L. cv. Great Lakes 366) was applied as the examination plant not rice. That is, the sample solutions were prepared refer to the weight of the original fruiting bodies. By the way, if specific activity is applied, most strong active compound, which is effective at lowest concentration, could be isolated. On the other hand, since total activity contains the factor of quantity, the most contributing compound to the total allelopathic activity of the mushroom.

The fruiting bodies were extracted with organic solvents in the general way, and were fractionated with several chromatographic procedures. The obtained extracts and chromatographic fractions were assessed allelopathic activity by the result of the biological assay, and two major allelochemicals were isolated. These purified compounds were identified as CPA-H and 15-hydroxy-CPA-H by spectroscopic analysis (Figure 5).

The growth inhibitory activity for the lettuce radicle of these allelochemicals was 0.62, 0.99 mM (IC_{50}) respectively. However, the allelopathy of the mushroom in nature was not supported by their intermediate active strength and the amount. Then, it was thought that active allelochemicals were supplied from less active precursors as same case as the black walnut mentioned above. After comparing the structure of dimeric CPAs which occupied 1-2% of the dried mushroom weight, it is found that all CPA could be converted to two isolated allelochemicals when all ester bonds were disconnected. The ester bond was the structure that was easy to be disconnected of structure of CPAs most, and it was thought that dimer CPAs developed allelopathy by conversion into monomers in the process when a fruit body was broken down on the ground. The same bioassay was performed about similar bioactive examination about thieteen kinds of dimeric CPAs which were prepared from fractions not showed significant activity in the allelochemical isolation process. As expected, no compound was detected appearing strong growth inhibitory activity as expected, and IC_{50} values of all thirteen CPAs were less than 100 mM and were not able to be calculated without being admitted. This result is different from the expectation from inhibitory activity in rice [35]. Then, an investigation which is inactive dimeric-type CPAs of the fruiting bodies in the soil turn over to active allelochemicals (CPA-H and 15-hydroxy-CPA-H) by the soil organisms, or not were performed. The peak of the inhibitory activity of the soil, which was crowded pulverized fruiting bodies, was observed at 4th day. The result shows that the potent allelochemical precursors were transformed by soil.

On the other hand, the amounts of existing active form CPAs in original material and active CPAs where it could supply by cleaving of ester bonds were quantified, and the allelopathy potential of the mushroom was estimated. As a consequence of aqueous sodium hydroxide treatment, active allelochemicals CPA-H and 15-hydroxy-CPA-H was increased drastically 40 times and 2,000 times respectively. Then, the total activity [36] of the mushroom was evaluated and compared with those of other allelochemicals. The obtained value 386 showed that the *C. volvatus* had very high allelopathy potential among known allelopathic plants to date.

Through these experiments, it was clarified that *C. volvatus* produced a large amount of dimeric CPAs as inactive form, but disconnection of ester bonds by the action such as microbes change the compounds into active monomeric CPAs (CPA-H and 15-hydroxy-CPA-H) to show allelopathy. However, the allelopathy of the mushroom has many more unexplained all facts. It is certain that the mycelium spreading through the neighborhood acts, but there has been not any evidence yet. The detail metabolism of CPAs degradation in nature has been not investigated yet. There is not any information of the mode of action of active CPAs. The author wants to elucidate it even little by little in future.

THE ALLELOCHEMICAL OF *BOLETUS FRATERNUS* FRUITING BODY [8]

Fruiting bodies of *Boletus fraternus*, an edible but indelicious mushroom, occurs in summer mainly in Japan. The author decided to clarify the allelochemical of the mushroom, because there were extremely few broad leaf weeds around the place on the author inspection. A fruiting body was extracted with organic solvents according to a usual procedure. The obtained extracts were chromatographed to purify an allelochemical with an assessment of biological assay using lettuce seedling, and an allelochemical was isolated. Each sample solutions were prepared on the basis of total activity [36] in same way of *C. volvatus*. As a result, 2(*S*)-amino-4(*R*)-methylhex-5-enoic acid (AMHA, Figure 6), non-protein amino acid, was identified by spectroscopic analysis and comparison with literature data. Interestingly, when the sample solution was at high density more than 100 ppm, remarkably growth inhibited radicle turned up against gravity to prevent from contact with filter papers including the allelochemical containing test liquid. This phenomenon suggests a possibility AMHA bend nearby plants root to have priority to plants. Knowledge is not provided about the mode of action at all. It is well-known that many mushrooms produce specific non-protein amino acids [37] such as above-mentioned ACPBA and ACPA, [21] and it is expected that these compounds show various physiological functions by competing with usual amino acids necessary for life maintenance.

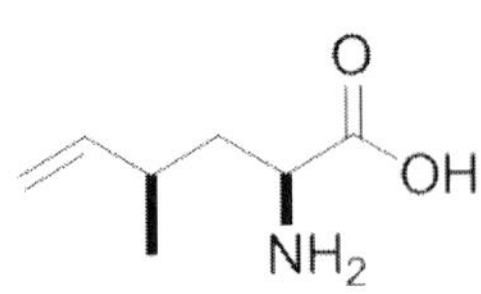

Figure 6. AMHA, an allelochemical, produced by *Boletus fraternus*.

AN ALLELOCHEMICAL PRODUCED BY *POLYPORUS ARCULARIUS* MYCELIUM [9]

Polyporus arcularius has been distributed over the world. The mushroom is not poisonous but unsuitable to eat because of their toughness. In some area in Japan, the mushroom dried has been used as a material to make bouillon stock. The author had found the occurrence of the fruiting bodies on a stump of Japanese apricot for several years. There are few weeds around the stump more clearly than other place. However, fruiting bodies of the quantity necessary for constitutional study were not obtained unfortunately. Therefore mycelium was isolated by an inoculation of a piece of internal tissue of the fruiting body to potato dextrose agar or Hamada medium plates. An allelochemical was isolated from the culture broth with same strategy of *B. fraternus* case, and the structure was determined to 2-hydroxyalbicanol (Figure 7) by analysis of spectroscopic data. The compound exhibited growth inhibition of lettuce seedling with IC_{50} values of 1.3 mM to hypocotyl and 1.7 mM to radicle. In the study, 2-hydroxyalbicanol was isolated from the culture broth not fruiting bodies. The fact indicates that the allelopathy was only showed only *in vitro*, and it's necessary that this compound is produced in wild fruiting body to link with the phenomenon.

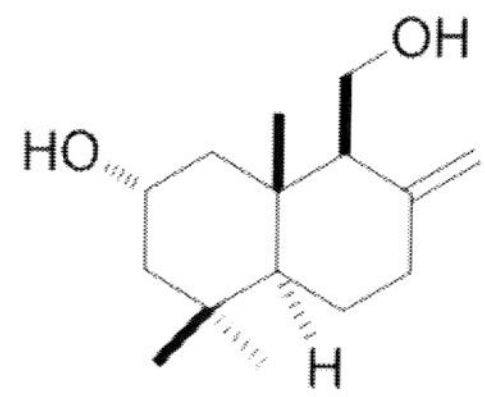

Figure 7. 2-Hydroxyalbicanol, an allelochemical isolated *Polyporus arcularius* culture broth.

Conclusion

In this chapter, the author reviewed the present status of mushroom allelopathy study with introduction of the author's work. The content is updated recent reviews written in Japanese [3]. The concluding remark is "all mushrooms show allelopathy" in this chapter.

Some questions appeared in the process when the author done the constitutional study of mushrooms. As an example, when mycelium of *C. volvatus* was cultured, CPA-D was only isolated and any other CPAs were not detected at all. Does the mushroom really produce CPAs except for CPA-D? It is said that the basidiomycete is hard to have the epigenetical metabolism control by substances from their circumstance. Does *C. volvatus* make various CPAs by the epigenetical control? Do symbiotic microbes and/or insects carry out the chemical conversion of CPA-D? The secondary metabolites produced at natural environments are more complicated than artificial culture condition at least, and it seems that these complicated metabolites make a more complicated chemical ecological phenomenon. It lets the mushroom allelopathy study complicate that difference may be seen in the constituents within same mushroom species but collected at different location. Wada, et al. reported that chemotype exist in *Boletinus cavipes*, [38] and the author confirmed that *Sarcodon scabrosus* has chemotype, too [6, 39]. The existence of chemotype may be going to lack in persuasive power to explain a certain genus mushroom allelopathy under natural environments.

Mushrooms produce specific interesting secondary metabolites including compounds introduced in this chapter. Still more a variety of compounds might be found in the near future by epigenetical metabolic controls, etc. The author aspire that the constitutional studies of mushrooms including the allelopathy study develop, and some of compounds are utilized for global ecosystem and our life wealthily in the future.

References

[1] Schmit, JP; Mueller, GM. An estimate of the lower limit of global fungal diversity. *Biodiv. Conserv.*, 16, 99-111 (2007).

[2] Zjawiony, JK. Biologically active compounds from Aphyllophorales (polypore) fungi. *J. Nat. Prod.*, 67, 300-310 (2004); Lindequist, U; Niedermeyer, TH; Jülich, WD. *eCAM*, 2, 285-299 (2005).; D. N. Quang, DN; Hashimoto, T; Asakawa, Y. Inedible mushrooms: a good source of biologically active substances. *Chem. Rec.*, 6, 79 (2006).; Zhong, JJ; Xiao, JH. Secondary metabolites from higher fungi: discovery, bioactivity, and bioproduction. *Adv. Biochem. Eng./Biotechnol.*, 113, 79 (2009).

[3] Araya, H. Secondary metabolites give influence to plant growth – does mushrooms show allelopathy? *Kinokokendayori* (The Mushroom Research Institute of Japan), 37, 19-28 (2014) (in Japanese); Araya, H. Mushroom of allelopathy. *Finechemicals* (CMC Publishing CO., LTD.), 44, 39-48 (2015) (in Japanese).

[4] Araya, H. Allelopathic acitivities in litters of mushrooms. In: Clark, JM. and Ohkawa, H. (eds.). *New Discoveries in Agrochemicals*. ACS, 63-72 (2005).

[5] Araya, H. Fruiting bodies of mushrooms as allelopathic plants. In: Fujii, Y. and Hiradate, S. (eds.). *Allelopathy – New Concepts and Methodology*. Science Publisher, 341-352 (2007).

[6] Endo, Y; Minowa, A; Kanamori, R; Araya, H. A rare α-pyrone from bitter tooth mushroom, *Sarcodon scabrosus* (Fr.) Karst. *Biochem. System. Ecol.*, 44, 286-288 (2012).

[7] Otaka, J; Taiko, S; Goseki, S; Miketa, M; Araya, H. Isolation and bioassay studies of Cryptoporic acids from *Cryptoporus volvatus* fruiting bodies. *J. Soc. Mushroom Sci. Biotechnol.*, 21, 113-122 (2013).

[8] Araya, H; Nagai, Y; Otaka, J. Isolation of (2*S*,4*R*)-2-amino-4-methyl-hex-5-enoic acid, a nonprotein amino acid, as an allelochemical from the fruiting bodies of *Boletus fraternus* Peck. *J. Plant Interact.*, 9, 627-631 (2014).

[9] Otaka, J; Araya, H. Two new isodrimene sesquiterpenes from the fungal culture broth of *Polyporus arcularius*. *Phytochemistry Lett.*, 6, 598-601 (2013).

[10] Araya, H; et al., The manuscript is under preparation.

[11] Molish, H. Der Einfluβ einer Pflanze auf die andere Allelopathy. Jena, Gustav Fischer (Germany) (1938). English translated edition: Ed. Narwal, SS. The Influence of one plant on another: Allelopathy. Jodhpur, Scientific Publishers (India) (2001).

[12] Chaumont, JP; Simeray, J. Propriétés Allélopathiques de 114 extraits de carpophores de champignons sur la germination de semences radis, *Rev. Écol. Biol. Sol.*, 22, 331-339 (1985).

[13] Uzun, Y; Türker, M; Kaya, A; Demirel, K; Işler, S; Keleş, A. Allelopathic potential of macrofungi on germinating maize (*Zea mays* L.) grain. *African J. Biotechnol.*, 9, 1016-1023 (2010).

[14] Mo, MH; Ma, HM; Xiao, QF. Study of the allelopathic effects of the ethanol-soluble extracts of *Lactarius hatsudake* on *Oryza sativa* and *Echinoloa crusgall. Acta Ecol. Sin.*, 24, 2951-2954 (2004).; Mo, M; Xiao, Q; Zhang, X; Nie, C. Allelopathy of aqueous extracts of *Lactarius hatsudake* on several crops and barnyardgrass (*Echinochloa crusgalli* L.). Fourth World Congress on Allelopathy, 21-26 August 2005, Charles Sturt University Wagga Wagga, NSW, Australia.

[15] Streiblová, E; Gryndlerová, H; Gryndlert, M. Truffle brûlé: an efficient fungal life strategy. *FEMS Microbiol. Ecol.*, 80, 1-8 (2012).

[16] Ikeda, M; Sato, Y; Sassa, T; Miura, Y. Structures of fasciculols, new plant growth inhibitors from *Naematoloma fasciculare. The 21st Symposium on the chemistry of natural products*, 21, 584-591 (1978).

[17] Choi, JH; Abe, N; Tanaka, H; Fushimi, K; Nishida, Y; Morita, A; Kiriiwa, Y; Motohashi, R; Hashizume, D; Koshino, H; Kawagishi, H. Plant-growth regulator, imidazole-4-carboxamide, produced by the fairy-ring-forming fungus *Lepista sordida. J. Agric. Food Chem.*, 58, 9956-9959 (2010).

[18] Choi, JH; Fushimi, K; Abe, N; Tanaka, H; Maeda, S; Morita, A; Hara, Y; Motohashi, R; Matsunaga, J; Eguchi, Y; Ishigaki, N; Hashizume, D; Koshino, H; Kawagishi, H. Disclosure of the "fairy" of fairy-ring-forming fungus *Lepista sordida. ChemBioChem*, 11, 1373-1377 (2010).

[19] Kawagishi, H. The chemistry of fairy ring and fairy chemicals – a possibility to plant growth regulators. *Kagaku to Seibutsu*, 52, 665 (2014).

[20] Mitchinson, A. Fairy chemicals. *Nature*, 505, 298 (2014).

[21] Yoshimura, H; Takegami, K; Doe, M; Yamashita, T; Shibata, K; Wakabayashi, K; Soga, K; Kamisaka, S. α-Amino acids from a mushroom, *Amanita castanopsidis Hongo*, with growth-inhibiting activity. *Phytochemistry*, 52, 25-27 (1999).

[22] Ikeda, M; Naganuma, Y; Ohta, K; Sassa, T; Miura Y; Isolation and Identification of a Plant-Growth Inhibitor, Azetidine-2-carboxylic Acid, from *Clavaria miyabeana* S. Ito and Its Occurrenence in the Family *Clavariaceae. Jpn Soc. for Biosci, Biotech. Agrochem.*, 51, 519-522 (1977).

[23] Fushimi, K; Anzai K; Tokuyama, S; Kiriiwa, Y; Matsumoto, N; Sekiya, A; Hashizume, D; Nagasawa, K; Hirai, H; Kawagishi, H. Agrocybynes A-E from the culture broth of *Agrocybe praecox. Tetrahedron*, 68, 1262-1265 (2012).

[24] Saito, A; Murayama, T; Nazarova, M; Ikeda, M. Study on the metabolites of *Suillus luteus* (produced in Russia). *Jpn Soc. for Biosci, Biotech. Agrochem*, 69, 235 (1995).

[25] Ikeda, M; Kanou, H; Nukina, M. Plant growth inhibitors from fruit bodies of *Polyporus confluens. Bull. Yamagata Univ. (Agric. Sci.)*, 10, 849-852 (1989).

[26] Shibata, M; Sakaki, T; Miyauchi, S; Wakamatsu, K; Mukai, K. A plant germination suppressor and its usage. JP-Patent 2005-281187.

[27] Okamoto, Y; Ohmae, M; Tanaka, S. A new compound, MU-118. JP-Patent 2001-302671.

[28] Sugasawa, H; Kurihara, N; Nazarova, M; Murayama, T; Ikeda, M. Plant growth regulating substances in *Stropharia aeruginosa* Q. *Jpn Soc. for Biosci, Biotech. Agrochem*, 70, 222 (1996).

[29] Fujii, Y. Allelopathy of velvetbean: determination and identification of L-DOPA as a candidate of allelopathic substances. In: HG. and Cutler, SJ. (eds.), *Biologically Active Natural Products*, 33-48, CRC Press (1999).

[30] Steglich, W; Esser, F. L-3,4-Dihydroxyphenylalanine aus *Strobilomyces floccopus. Phytochemistry*, 12, 1817 (1973).

[31] Steglich, W; Preuss, R. L-3,4-Dihydroxyphenylalanine from carpophores of *Hygrocybe conica* and *H. ovina. Phytochemistry*, 14, 1119 (1975).

[32] Hiradate, S. Isolation strategies for finding bioactive compounds: specific activity vs. total activity. In: Rimando, AM. and Duke, SO. (eds.), *Natural Products for Pest Management*, ACS Symposium Series 927, 113-126 (2006).

[33] Hatanaka, S. Amino acids from mushrooms, In: Herz, W; Kirby, GW; Moore, RE; Steglich, W; Tamm, Ch. (eds.), *Progress in The Chemistry of Organic Natural Products*, 59, 1-140, Springer-Verlag (1992).

[34] Otaka, J. Study on allelopathy of basidiomycetes *Cryptoporus volvatus* and *Polyporus arcularius*. Ph. D Thesis, Meiji University, Tokyo, Japan (2014).

[35] Wu, SH; Zang, M. *Cryptoporus sinensis* sp. Nov, a new Polypore found in China. *Mycotaxon*, 74, 415-422 (2000).

[36] Kato-Noguchi, H; Fushimi, Y; Kimura, F; Morita, M; Suenaga, K. Organ-specific-active allelopathic substance in red pine needles. *Plant Growth Regul.*, 68, 171-175 (2012).

[37] Asakawa, Y; Hashimoto, T; Mizuno, Y; Tori, M; Fukazawa, Y. Cryptoporic acids A-G, drimane-type sesquiterpenoid ethers of isocitric acid from the fungus *Cryptoporus volvatus*. *Phytochemistry*, 31, 579-592 (1992).

[38] Wada, T; Kobata, K; Hayashi, Y. Shibata, Y. Two Chemotypes of *Boletinus cavipes*. *Biosci. Biotech. Biochem*, 59, 1036-1039 (1995).

[39] Masubuti, H; Endo, Y; Araya, H; Uekusa, H; Fujimoto, Y. Establishment of benzodioxazine core structure for sarcodonin class of natural products by X-ray analysis. *Org. Lett*, 15, 2076-2079 (2013).

In: New Developments in Allelopathy Research
Editor: Julia E. Price

ISBN: 978-1-63483-390-5

Chapter 2

IS GROWING BUCKWHEAT ALLELOPATHIC?

Aurélie Gfeller and Judith Wirth*
Agroscope, Institute for Plant Production Sciences IPS, Research division: Plant protection for field crops and viticulture/oenology, Research group: Herbology in field crops and viticulture, Nyon, Switzerland

ABSTRACT

In a growth chamber trial the suppressive potential of common buckwheat (*Fagopyrum esculentum* Moench) on redroot pigweed (*Amaranthus retroflexus* L.) was tested by separating resource competition for light, water and nutrients from belowground root interactions between the two species. The growth repressing effect of buckwheat on redroot pigweed was very strong (> 89%) and due to both competitive shading effects and root interactions of potentially allelopathic nature [1]. In order to verify this result in situ a field trial was performed in 2014. Buckwheat and redroot pigweed were sown together on the 7th of August and the dry weight of redroot pigweed was determined 28 and 55 days after sowing (DAS). A strong repression of redroot pigweed growth by buckwheat could be observed at both dates independently of shading. When soil samples from the field within the buckwheat stands 11, 26 and 42 DAS were transferred into the growth chamber no effect on redroot pigweed and lettuce (*Lactuca sativa*) growth could be observed. Assuming that allelopathic molecules must be present in the soil solution supplementary experiments were conducted in the phytotron. Under controlled conditions buckwheat was grown in pots for different periods of time. These pots were used for several growth experiments with lettuce. Lettuce root length was measured after exposing seeds to

- fresh, dry soil from the pots,
- soil solution samples collected with Rhizon soil moisture samplers and
- boiling water extracts of the soil.

Moreover, buckwheat and lettuce were sown at the same time in petri dishes in soil and on blotting paper; subsequently lettuce root length was measured after 3 days. In none of the experiments an influence on lettuce and redroot pigweed development could

* E-mail: judith.wirth@agroscope.admin.ch.

be observed. In addition, redroot pigweed and buckwheat roots were *separated with a permeable membrane in a pot trial allowing molecules to diffuse between the compartments for 28 d. Redroot pigweed grown in the membrane was of the same size than in the control without buckwheat.* We therefore conclude that there

- are either no allelopathic molecules in the soil solution (not soluble in water) or that they are rapidly degraded,
- roots must be in direct interaction to allow allelopathic effects,
- the growth inhibiting effect is due to a long term and constant exposure of small quantities of allelopathic molecules or
- the growth repression of redroot pigweed by buckwheat is not due to allelopathy.

INTRODUCTION

Rice, barley, wheat, rye, sorghum and sunflower are known to be allelopathic crops [2, 3] and a lot of research has been done on them. Weed suppressive rice cultivars are now commercially available in the U.S. and China [4]. For cover crops allelopathy has only been shown for a few species, like for example hairy vetch (*Vicia villosa*) and oat (*Avena sativa*) [2, 5, 6]. Common buckwheat (*Fagopyrum esculentum*) is also known to successfully suppress weeds during its growth in the field [7-12] and after residue incorporation [4, 13-17]. For centuries, farmers have been using its empirically known ability to control weeds [11, 18, 19] and its cultivation is used for weed suppression in crop rotations and traditional agricultural systems due to its short life cycle and fast decomposition [20, 21].

However, it is not clear what the weed suppressive ability of this cover crop is due to, although several articles claim that it is due to allelopathy [11, 19, 22-25]. In a recent review it was described that several factors might explain the weed suppressive potential of growing buckwheat stands and that allelopathic root exudates are one possibility to explain the phenomenon [8]. In the following text we want to explain how we tried to find out whether phytotoxic root exudates are present in soil in which buckwheat is growing and why it is important to choose the appropriate method to study allelopathic root interactions. For our experiments we used different methods that have been published before in this field of research. In order to determine allelopathic activity of root exudates the potentially allelopathic plant (donor) and the targeted plant species (receiver) can be co-cultivated in petri dishes. After several days of co-culture length and/or weight of hypocotyls and roots of the receiver plant are determined [19, 24, 26]. Another possibility is to use soil in which the donor plant has been growing for cultivation trials with the receiver plant [26-28]. The soil can also be extracted with different solvents. Some authors propose water as extraction medium [23, 29, 30]. With our own experimental design we are trying to separate resource competition for light, nutrients, water and space from allelopathic root exudates present in the soil [1].

It is known that organic toxins produced by an allelopathic species can be rendered harmless by the combined interactions of soil texture, organic matter, temperature, irradiance and microbial breakdown and that therefore a laboratory bioassay alone does not adequately demonstrate that allelopathy is operational in the field [31]. In order to influence growth of plants in field situations, allelochemicals must accumulate and persist at phytotoxic levels and come in contact with the target plant [32]. For this reason we performed field trials and bioassays with non-sterile field soil. In our opinion it is very important to work with growing plants in living soil as it has been shown that changes in soil microbial community induced by root exudates can be responsible for growth suppression [32, 33].

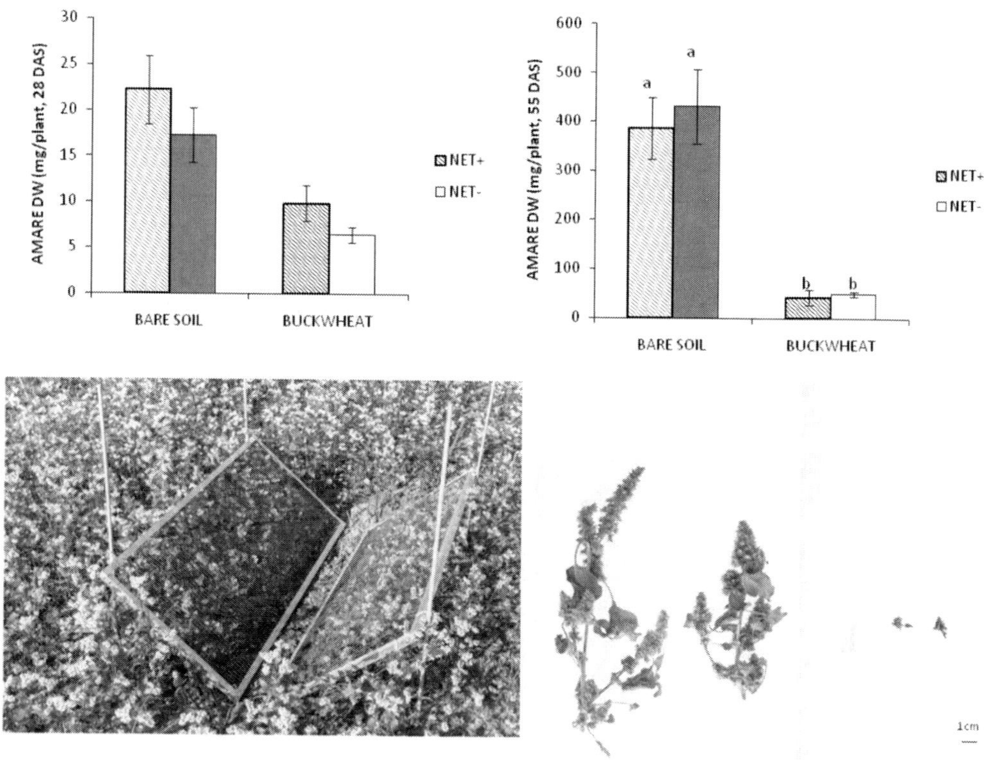

Figure 1. Growth repression of AMARE by buckwheat in the field.
AMARE DW was determined after 28 d (A) and 55 d (B) of growth in a buckwheat field (BUCKWHEAT) and in control plots with bare soil (BARE SOIL) in the presence (NET+) or absence (NET-) of nets preventing shading. Values are means ± SE of 8 replicates for buckwheat and 16 for bare soil. *Means* not sharing the same letter are *significantly different (Tukey's HSD, $p < 0.05$)*. Picture of the experimental design (buckwheat, Net+) in the field (C). AMARE harvested after 55 d of growth on bare soil (D) or in a buckwheat stand (E).

Material and Methods

After field preparation with a rotary harrow, buckwheat was sown on the 7th of August 2014 at a density of 75 kg/ha and a row spacing of 12.5 cm on a loamy soil (sand 30%, clay 25%, silt 46%, pH 6.6) close to the lake of Geneva in Switzerland. The preceding crop was winter wheat. Buckwheat was grown in four blocks of 24 m^2 each. In addition eight blocks of the same size were left with bare soil. Within each block two zones of 1 m^2 were defined. Within each zone one central inter-row of buckwheat was chosen for seeding of *Amaranthus retroflexus* (AMARE). AMARE seeds were purchased from Herbiseed (Twyford, England) and buckwheat seeds from OH semences (Switzerland).Within the inter-row, 120 AMARE seeds, previously thermised at 50°C for 1 hour (h), were sown at the same day as buckwheat in four small areas of about 10 cm, separated from each other by about 15 cm. The sowing areas in the inter-rows were subsequently covered with a voile for 7 days (d) in order to promote germination. Additionally around 1200 AMARE seeds were sown within the 1 m^2 zones (outside of the central inter-rows) in order to increase weed pressure. In one of the 1 m^2 zones of each block a net (1.2 m length x 0.5 m large, 12 mm mesh) was placed 11 days after sowing (DAS) in order to prevent shading of AMARE by buckwheat (Figure 1 C). 18 DAS the number of AMARE plants was reduced to 5 in each small sowing area. Throughout the experiment weeds, other than AMARE, were removed by hand within the 1 m^2 zones. Anti-slug granules (Capito, Landi, CH) were applied 22 DAS at a concentration of 70-80 granules per m^2. 28 and 55 DAS five AMARE plants of two small sowing areas per inter-row were harvested. Subsequently plant material was dried at 50°C for 5 d in order to determine dry weight (DW).

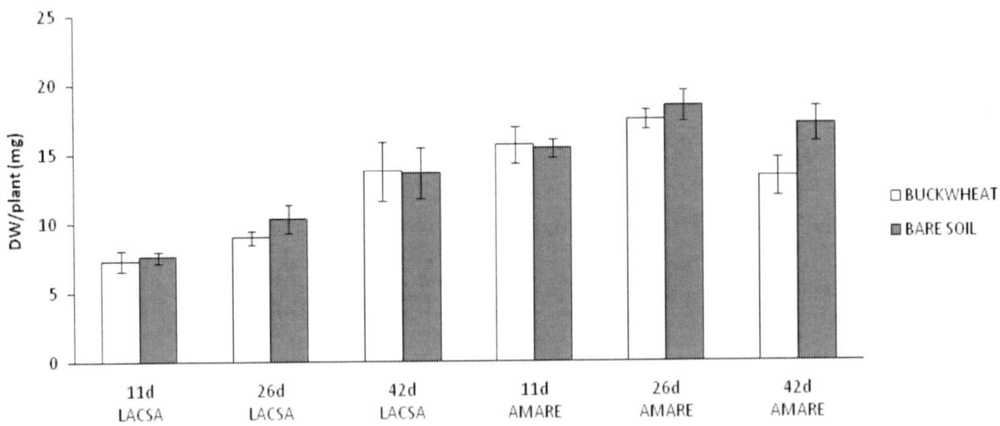

Figure 2. No effect of soil from a buckwheat stand on lettuce and AMARE growth.
Lettuce (LACSA) and AMARE (AMARE) were grown for 15 and 21 d respectively in a phytotron on soil taken from a buckwheat stand (BUCKWHEAT) or on bare field soil (BARE SOIL) 11, 26 and 42 d after buckwheat sowing. Values are means ± SE of 12 replicates for buckwheat and 24 for bare soil. *There were no statistically significant differences in DW between the samples from a buckwheat stand and from bare soil at the different dates (non-parametric Dunnett's test, $p < 0.05$).*

In the 2014 field trial, both under the buckwheat cover and on bare soil, 11, 26 and 42 DAS three soil samples (about 1 liter of soil, to a depth of about 15 cm) per block were taken. Each soil sample was separated and filled in two pots in which 10 lettuce (purchased from Bigler, Switzerland) or 30 AMARE seeds were sown. Pots were placed in a phytotron (for lettuce: Night (N)/Day (D) 10 h/14 h, Temperature 18/25°C, relative humidity (RH): 70%; for AMARE: N/D 10 h/14 h, 25/30°C, RH: 70%) and watered daily. After one week, the number of lettuce and AMARE plants was reduced to 5 plants per pot. Lettuce and AMARE plants were harvested after 15 and 21 d respectively and dried at 50°C for 72 h to determine DW (Figure 2).

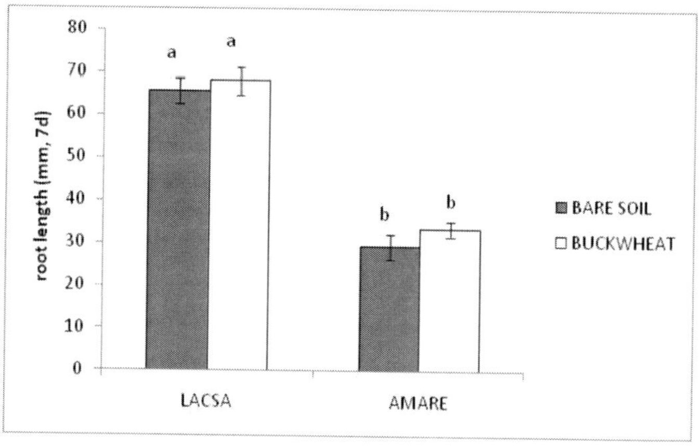

Figure 3. No effect of soil in which buckwheat was grown on lettuce and AMARE growth.
50 g of dried bare soil (BARE SOIL) and soil in which buckwheat (BUCKWHEAT) was grown for 10 d was transferred into petri dishes. 10 Lettuce (LACSA) or amaranth (AMARE) seeds were sown in each petri dish and root length was determined after 7 d. Values are means ± SE of 60 replicates for LACSA and 30 for AMARE. *Means* not sharing the same letter are *significantly different* (ANOVA, $p < 0.05$).

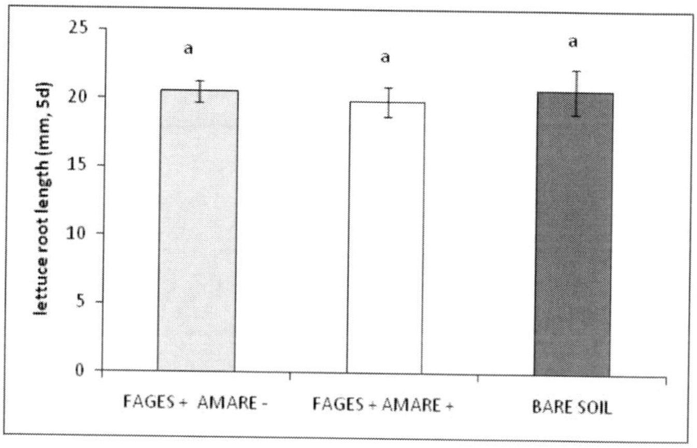

Figure 4. No effect on lettuce growth of soil water extracts obtained with rhizon samplers.
Soil extracts were collected with rhizon samplers from pots cultivated with buckwheat alone (FAGES+, AMARE-), buckwheat and AMARE (FAGES+, AMARE+) and bare soil (BARE SOIL) 28 DAS. 6 ml of each extract was applied on blotting paper on which lettuce was grown for 5 d in order to determine lettuce root length. Values are means ± SE of 30 replicates and 10 for bare soil. *There were no statistically significant differences* (ANOVA, $p < 0.05$).

Buckwheat was grown in pots (0.8 l) in the phytotron (N/D 10 h/14 h, 18/25°C, RH: 70%). The soil used in this experiment, as well as in all other experiments described further down, was non-sterile loamy field soil without any fertilization (26.5% clay, 41.9% silt, 31.6% sand, 2.8% organic matter, pH 7.9). Ten DAS the soil was taken, the roots were removed and it was dried for 1d at room temperature. Two blotting papers were put in a petri dish (10 x 10 cm), 10 ml of water was added and 10 lettuce or 10 AMARE seeds, previously thermised at 50°C for 1 h, were sown and covered with 50 g of dry soil. 20 ml of water was added to moisten the soil. Petri dishes were sealed and placed in a climate cabinet for 7 d

(N/D 10 h/14 h, 18/25°C) before measuring root length (Figure 3). The trial was repeated once with lettuce with the same results (not shown).

Buckwheat (100 seeds/pot) was cultivated in pots (5 l) in the phytotron (N/D 10 h/14 h, 18/25°C, RH: 70%). One pot of bare soil served as control. One Flex Rhizon sampler (Rhizosphere Research Products, Wageningen, The Netherlands) was introduced into each pot horizontally through a hole made 5 cm from the bottom of the pot at the day of sowing. 29 DAS, soil water samples were taken. Pots were irrigated with 1 l of water. When the water had stopped draining, soil water was obtained with the rhizon samplers. 6 ml of soil water from each pot was used to moisten a blotting paper in a petri dish. 10 lettuce seeds were sown on each paper in a line 1.5 cm below the top of the petri dish. Petri dishes were sealed and placed vertically in a climate cabinet for 5 d (N/D 10 h/14 h, 18/25°C). Subsequently lettuce root length was measured (Figure 4). The trial was repeated at 29, 36 and 43 DAS of buckwheat culture with the same results (not shown).

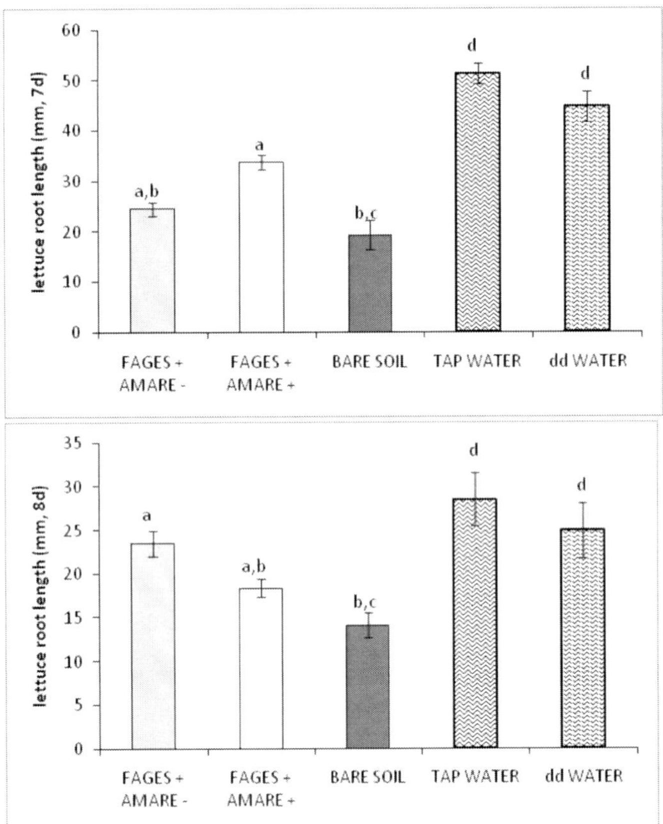

Figure 5. No effect on lettuce growth of water extracts from soil in which buckwheat was cultivated. Buckwheat alone (FAGES+, AMARE-) or buckwheat together with AMARE (FAGES+, AMARE+) were cultivated for 13 d (A) and 20 d (B). Control pots with soil only (BARE SOIL) were kept in the phytotron at the same time. The different soils were extracted with boiling water. 6 ml of extract were applied on blotting paper on which lettuce was grown for 7 in the light (A) or 8 d in the dark (B) in order to determine lettuce root length. Tap water (TAP WATER) and double distilled water (dd WATER) served as control. Values are means ± SE of 60 replicates for soil with buckwheat, 20 for bare soil and 10 for water controls. *Means* not sharing the same letter are *significantly different* (rank based analysis of linear models, $p < 0.05$).

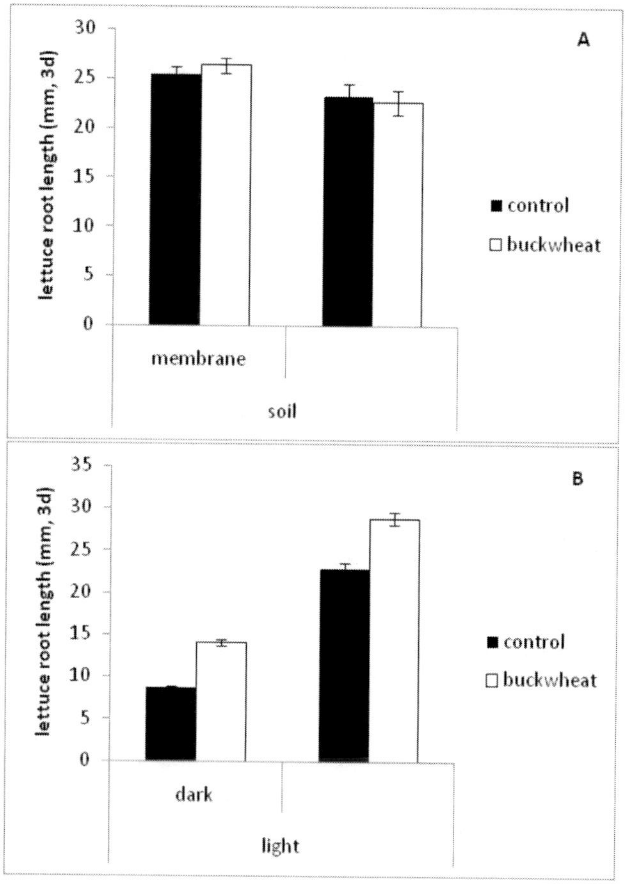

Figure 6. No effect of germinating buckwheat on lettuce root growth.
Lettuce was grown alone (control) or together with buckwheat (buckwheat) in petri dishes in a phytotron. The two species were growing in soil allowing root contact (A, soil) or they were separated by a permeable membrane (A, membrane). *There were no statistically significant differences in root length between the samples in the two different growth conditions (two-tailed paired student's t-test, $p < 0.05$).* In a second experiment the two species were growing on blotting paper under light conditions (B, light) or in the dark (B, dark). Lettuce root length was significantly longer when grown together with buckwheat *(two-tailed paired student's t-test, $p < 0.05$).* In both experiments (A and B) lettuce root length was determined after 3 d. Values are means ± SE of 50 replicates.

Buckwheat and buckwheat with AMARE were each grown in 3 pots (3 l) in the phytotron for 13 d and 20 d respectively (N/D 10 h/14 h, 18/24°C, RH: 70%). In addition 1 pot of bare soil was kept in parallel and watered two times at 0 d and 2 d before soil extraction. For the soil extraction 30 ml of soil were taken from the upper part of each pot and mixed with 20 ml of boiling tap water. Subsequently the samples were sonicated for 10 min before being put into a 60°C water bath for 30 min. After centrifugation (3500 rpm, 10 min), the supernatant was harvested and 2 times 6 ml of each pot was used to moisten 2 blotting papers in 2 petri dishes. 10 lettuce seeds were sown on each blotting paper. Petri dishes were sealed and placed vertically in a climate cabinet for 7 in the light or 8 d in the dark (N/D 10 h/14 h, 18/21°C). Subsequently lettuce root length was measured. Tap water and double distilled water were used as a control (Figure 5). The trial was repeated twice at 13, 20 and 28 DAS of buckwheat culture with the same results (not shown).

50 g of soil were sieved (< 3 mm) and evenly spread in squared petri dishes. 6 pre-germinated buckwheat seeds (4 d in the dark at 18/24°C, N/D 10 h/14 h) together with 10 lettuce seeds were placed on the soil (Figure 6 A, soil). Lettuce seeds without buckwheat served as control. In order to verify whether root interactions between buckwheat and lettuce are necessary for the release of allelochemicals, 30 g of soil were put in petri dishes. 6 pre-germinated buckwheat seeds were placed into the soil. Subsequently the soil was covered with a membrane (Sefar Nitex, 30 micron) of the size of the petri dish. On top of the membrane another 20 g of soil was placed and 10 lettuce seeds were sown (Figure 6 A, membrane). 30 ml of tap water was added to all petri dishes. Petri dishes were sealed and placed in a phytotron for 3 d (N/D 10h/14 h, 18/24°C). Subsequently lettuce root length was measured. The trial was repeated once with the same results (not shown).

6 pre-germinated buckwheat seeds together with 10 lettuce seeds were placed on blotting paper in petri dishes. In the control treatments only 10 lettuce seeds were placed in each petri dish. 15 ml of tap water was added to all petri dishes. Petri dishes were sealed and placed in a phytotron for 3 d (N/D 10h/14h, 18/24°C). One half of the petri dishes was incubated in the dark (Figure 6 B, dark) and the other half in the light (Figure 6 B, light). The trial was repeated once with the same results (not shown).

In a pot trial in the phytotron we recently showed that AMARE growth suppression by buckwheat is due to shading effects and root interactions of potentially allelopathic nature [1]. In order to exclude root competition for space and to test whether AMARE growth suppression is due to allelopathic molecules that diffuse in the soil solution a supplementary experimental condition was added. A permeable membrane (Sefar Nitex, 30 micron) was used to separate the roots from the two species. In the two other experimental conditions roots were allowed to interact or separated by an impermeable plastic barrier as described by Falquet et al. (2014). In all pots nets were installed to prevent shading of AMARE by buckwheat 11 DAS. Nutrient level and water supply were constantly maintained at a sufficient level. All pots were placed in a phytotron (N/D 10h/14h, 18/24°C, RH: 70%). Six different growth conditions were tested (Figure 7): i) buckwheat and AMARE, no root separation (= 0); ii) AMARE alone, no root separation (= 0); iii) buckwheat and AMARE, root separation by permeable membrane (= M); iv) AMARE alone, root separation by permeable membrane (= M); v) buckwheat and AMARE, root separation by plastic barrier (= P); vi) AMARE alone, root separation by plastic barrier (= P). Buckwheat seeds (cv 'Lileja') were purchased from UFA Samen (Switzerland). On day 1 of the experiment 8 pregerminated buckwheat seeds were sown in each external compartment and 30 AMARE seeds were sown in the central compartments of the pots. 8 DAS AMARE plants were thinned to five plants per pot based on a uniform distribution in the central compartment of each pot in order to avoid intraspecific competition. 28 DAS AMARE plants were harvested and dried for 48 h at 50°C to determine DW.

Results and Discussion

Due to its very rapid growth, it has been proposed that the competition for light is responsible for the major part of weed suppression by buckwheat [34], as a thick canopy prevents weeds from receiving enough light and therefore acts as a good competitor. In our own field trial in 2013 we could demonstrate that shading played a decisive role in AMARE growth suppression by buckwheat [12]. A partial contribution of shading in AMARE growth suppression by buckwheat could also be shown in our phytotron pot trials [1]. The total growth repressing effect of buckwheat on AMARE was very strong in two independent experiments (89% and 93% respectively), however it could only partially be explained by shading (68% and 65% respectively). In order to retest AMARE growth suppression by buckwheat under natural conditions another field trial was performed in 2014. 28 DAS AMARE plants under the canopy of the buckwheat crop were significantly smaller (63%) than AMARE plants in bare soil (Figure 1 A, net-). AMARE plants within the buckwheat stand in the absence of shade (between the nets, Figure 1 C) were 55% smaller than control plants on bare soil (Figure 1 A, net+). No effect of shading could be observed. 55 DAS the growth repression of AMARE within the buckwheat stand, both in the absence and presence of shading, was of 89% and significant (Figure 1 B, D and E). It can be concluded from this experiment that under the field conditions of 2014 the growth repressing effect of buckwheat on AMARE was strong and independent of shading. In 2014 the average temperature over four weeks after sowing was of 17.0°C and rather cold (compared to 18.9°C in 2013). Moreover during the 3 week period after sowing the daily minimum temperature recorded was below 11°C on 8 d, whereas in 2013 a minimum daily temperature below 11°C only occurred once. Due to this cold weather period in august 2014 buckwheat developed very slowly and couldn't establish a very dense canopy. Light (PAR) measurements showed that 30 and 48% of light was intercepted by the buckwheat canopy at 26 and 48 DAS respectively (results not shown) whereas in 2013 light suppression by buckwheat was 85 and 98% at 25 and 45 DAS respectively [12]. Light competition by shading was therefore much smaller in 2014 as compared to 2013. However, shading within the buckwheat stands was present, but didn't cause growth suppression of AMARE. We therefore conclude that the root interactions between buckwheat and AMARE were responsible for the observed growth repression. Our hypothesis is that weed suppression by buckwheat is due to competition for light and/or allelopathic root interactions. Zhang et al. [35] claim that the weed suppression effect from a crop is a joint action of allelopathy and competitive factors. Depending on the growth conditions of buckwheat in different environmental conditions, one mechanism for weed suppression might be predominant. If buckwheat is able to rapidly develop a dense canopy there is less need for the production of allelochemicals, whereas if shading ability is low other mechanisms of weed suppression might be induced. According to Wang et al. [36] *Chenopodium acuminatum* plants could acquire sufficient resources to grow fast enough to avoid being shaded by the surrounding vegetation in a field trial, when the neighbors grew relatively slowly. This example shows that the environmental conditions strongly influence plant growth. As under our field conditions no fertilizer was applied we cannot exclude nutrient competition for growth repression, especially after 55 d. However, growth repressing effects might be due to allelopathic molecules exudated by the cover crop.

In order to test this hypotheses we took soil samples within the buckwheat stand and in the bare soil control plots at three different dates (11, 26 and 42 DAS). This method has been used to demonstrate allelopathic activity of an allelopathic rice cultivar [27]. When lettuce and amaranth were grown on this soil in a phytotron no differences in the DWs of the two tests plants could be observed at the end of the trial (Figure 2). The differences in lettuce DW between 11, 26 and 42 d were probably due to the use of different phytotrons. A similar experiment was performed by Kumar et al. [37]. Soil in which buckwheat had been growing for 35 d in the greenhouse was used for growth experiments with Powell amaranth (*Amaranthus powellii*) over 20 d. In buckwheat grown soil amaranth DW was reduced by 76 and 77% respectively in two subsequent years. The pots received no fertilization during buckwheat growth and nutrient deficiency in the buckwheat amended soil was very likely the cause of growth suppression. However, the presence of allelopathic molecules cannot be excluded. In our own experiment we did not make the same observations as Kumar et al. We suppose that possible present allelopathic molecules were not sufficiently concentrated or not persistent enough to provoke growth repression of tests plants under our trial conditions. As lettuce and AMARE DW on bare soil and buckwheat field soil 42 DAS (Figure 2) were the same, we conclude that the nutrient content was similar between the two soil types. Therefore we assume that the strong AMARE growth repression observed in the field at 28 and 55 DAS (Figure 1 A and B) was not due to nutrient deficiency. As explained above, the buckwheat cover developed very slowly in 2014 and therefore probably didn't deplete nutrient resources.

Hayashi et al. [18] showed that intact common buckwheat plants didn't suppress the above ground plant parts of different weeds but only their radicle length. A stronger reduction of *Lolium rigidum* root growth compared to shoot growth was observed in studies of different allelopathic canola genotypes [38]. We therefore also studied the effect of "buckwheat soil" on radicle growth of lettuce. According to Kalinova et al. [19] dry soil from a buckwheat stand one day before harvest (130 DAS) had a growth inhibiting effect on lettuce roots after 3 d in the dark (-24%). The researches repeated the same experiment in another study at the stages of branching, flowering and before harvest [23]. The growth repressive effect of the dry soil before harvest could not be confirmed and the results were contradictory to the above mentioned publication. However, this time significant effects could be observed at the stages of branching and flowering. We repeated the same experiment with dry soil obtained from a pot trial in the growth chamber (10 DAS). Lettuce and AMARE root length was determined after 7 d of growth. No effect on the root length of the two species could be observed (Figure 3). We conclude that methods using "buckwheat-grown soil" lead to contradictory results and are not reliable for the study of allelopathic buckwheat root exudates. Moreover, it has been shown that the activity of an allelopathic compound can vary considerably between different soil types [39].

Subsequently we wanted to test whether allelopathic molecules can be found in the soil solution of pots in which buckwheat was grown. When we took soil water samples with rhizon samplers for germination tests with lettuce we could see no difference between pots in which buckwheat (± AMARE) was grown and bare soil (Figure 4). We therefore conclude that allelopathic molecules were present in very low concentrations, that they were rapidly degraded in the climate cabinets during the 5 day lettuce growth test or that buckwheat phytotoxins are not soluble in water. According to Duke [40] there are good scientific reasons to hypothesize that the most effective allelochemicals would have very limited water solubility. Sorgoleone is an example of a lipophilic allelochemical that would not be found by only aqueous extraction of the plant [41]. Moreover it was shown that water-soluble forms of different phenolic compounds accounted for less than 0.7% of the total estimated by extraction of the soil with NaOH [42]. Kalinova et al. [19] also showed that soil samples from a buckwheat stand extracted with distilled water showed no effect on lettuce root and hypocotyl elongation after 3 d. Tominaga et al. [11] showed that water extracts from field soil in which buckwheat was grown had a growth repressive effect on root length of *Echinochloa crus-galli* and *Portucala oleracea* after 3 d. However, growth repressive effects were not very strong, standard errors were not presented and no statistical analysis was performed. In order to verify our results we tested another method. Soil in which buckwheat (± AMARE) was grown was extracted with boiling water according to Kalinova et al. [23] as more/other molecules might be liberated during the extraction process. However, in our experiments boiling water extracts from bare soil showed strongest growth suppression of lettuce, both after 13 and 20 d of buckwheat growth (Figure 5 A and B). We cannot explain this result. When tap water or double distilled water were taken as control lettuce root length was significantly increased compared to the water extracts from buckwheat containing soil (Figure 5 A and B). Looking at these results (without considering the control on bare soil) one could conclude that water extracts from buckwheat soil contained growth repressing substances. We therefore think that this type of essay is inappropriate to study the allelopathic potential of buckwheat.

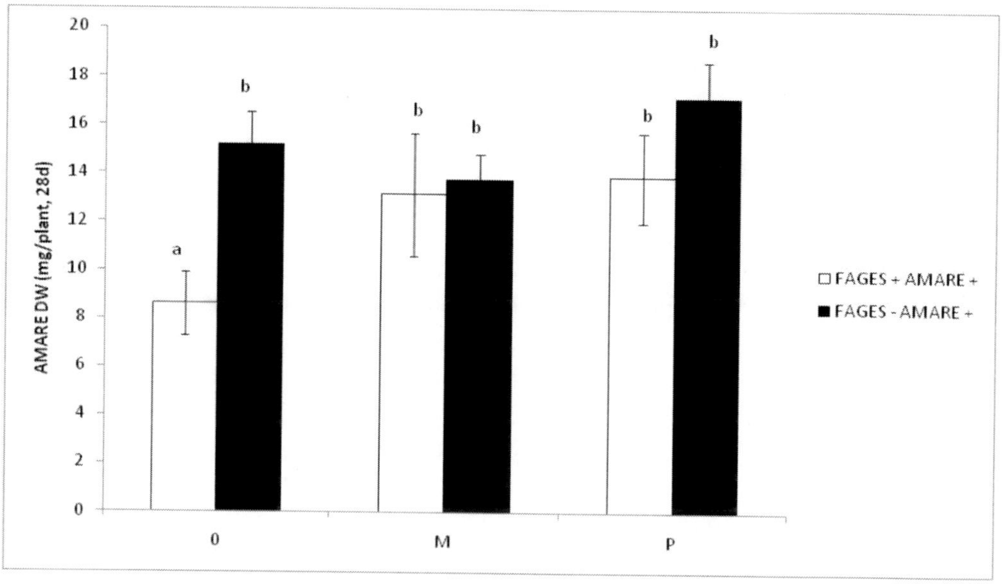

Figure 7. Growth repression of AMARE by root interactions with buckwheat.
AMARE with buckwheat (FAGES+, AMARE+) and AMARE alone (FAGES-, AMARE+) were grown for 28 d in pots in a system preventing shading of AMARE by buckwheat via nets. AMARE and buckwheat roots could interact (0), or were separated by a permeable membrane (M) or an impermeable plastic barrier (P). AMARE DW was determined after 28 d. Values are means ± SE of 6 replicates. *Means not sharing the same letter are significantly different (Tukey's HSD, $p < 0.05$).*

According to Kalinova et al. [19], during germination and early development of buckwheat several allelopathic compounds are exudated. The authors showed that lettuce radicle length was suppressed by 36% after 4 d in the dark when grown next to germinating achenes of buckwheat. Kato-Noguchi et al. [24] cultivated buckwheat seedlings together with different receiver plants (including lettuce) for 3 d at 25°C with a 12 h photoperiod in a growth chamber. A strong reduction of lettuce root growth (80%) could be observed. The authors could also show that the growth inhibiting effect was stronger when the number of buckwheat seedlings was bigger. In our own experiments, when lettuce and young seedlings of buckwheat were grown together in petri dishes for 3 d no growth repressing effect on lettuce root length could be observed, neither in soil (Figure 6 A) nor on blotting paper

(Figure 6 B). We could not confirm the growth repressive effect of germinating buckwheat seedlings on lettuce. On the contrary, we found a stimulating effect of young growing buckwheat seedlings on lettuce in the absence of soil both under light conditions (Figure 6 B, light) and in the dark (Figure 6 B, dark). However, we most probably didn't use the same buckwheat cultivars. It is possible that the cultivars used in the two articles mentioned above had a stronger growth repressive potential. With these experiments we could also show that it doesn't make a difference whether the roots of the two species are touching.

Based on our protocol published in 2014 [1], the method of separating growth repressive factors was developed and improved. In addition to an impermeable plastic barrier in between the roots of the two species a permeable membrane was chosen as an additional experimental condition. The membrane allows water, nutrients and hydrophilic molecules to circulate between the compartments. In the presented trial nets for shade avoidance were placed in all pots in order to exclude light competition and fertilization was kept on a high level throughout the trial. In a recent article the effects of competition and fertilization on *Chenopodium acuminatum* were studied. It was shown that root competition was stronger than shoot competition without fertilizer and that the addition of fertilizer strongly reduced the effects of root competition [36]. We therefore assume that plants didn't compete for nutrients in our trial conditions, as it is to be expected that addition of a resource will reduce competition for that resource [36]. It can be seen that amaranth growth was significantly repressed when the roots of the two species interacted (Figure 7, condition O). However, inside the permeable membrane amaranth growth was not reduced. We therefore hypothesize that potentially allelopathic molecules did not diffuse within the membrane. They could either be not hydrophilic or the distance between the roots of the different species was too large. We conclude that buckwheat and amaranth roots might have to touch in order to provoke growth repression of amaranth. The important role of root-root interactions on weed growth inhibition by allelopathic wheat has recently been shown by Zhang et al. [35]. Compared to the experiment presented in Figure 6 A, where we could not observe a growth repression of lettuce roots, we hypothesize that roots have to touch over a long period of time and/or that buckwheat has to be at an advanced growth stage in order to cause growth suppression. Moreover the effect seems to be cumulative as visually growth repression could only be observed during the last week of the four week trial (results not shown). It is generally accepted that allelochemicals are weak phytotoxins that exert their effects at low, but constant or increasing concentrations over long periods [41]. This corresponds to our observations. It is also possible that the presence of AMARE might induce phyotoxic root exudation of buckwheat. This phenomenon has been shown for an allelopathic rice variety that exuded significantly more phytotoxins from its roots when grown together with *Echinocloa crus-galli* [43].

CONCLUSION

With our experiments we could not prove that growing buckwheat secretes allelopathic root exudates. Our observations can be explained by different hypothesis: allelopathic compounds are not water-soluble, they are rapidly degraded after exudation or their concentration in the soil solution is very low. Another assumption is that the growth repression of redroot pigweed by buckwheat is not due to allelopathy. However, we have indications that roots of buckwheat and amaranth have to be in contact over a long period of time to trigger the release of allelochemicals and that the growth repressing effect is cumulative. Moreover we could show that light competition only plays a decisive role under certain growth conditions and that nutrient competition can probably be excluded as growth repressing factor.

REFERENCES

[1] Falquet, B., et al., Simple method to separate resource competition from allelopathic root interactions. *Allelopathy Journal,* 2014. 34(2): p. 227-240.

[2] Jabran, K., et al., Allelopathy for weed control in agricultural systems. *Crop Protection,* 2015. 72: p. 57-65.

[3] Belz, R.G., Allelopathy in crop/weed interactions - an update. *Pest Management Science,* 2007. 63(4): p. 308-326.

[4] Worthington, M. and C. Reberg-Horton, Breeding Cereal Crops for Enhanced Weed Suppression: Optimizing Allelopathy and Competitive Ability. *Journal of Chemical Ecology,* 2013. 39(2): p. 213-231.

[5] Campiglia, E., et al., Effect of cover crops and mulches on weed control and nitrogen fertilization in tomato (Lycopersicon esculentum Mill.). *Crop Protection,* 2010. 29(4): p. 354-363.

[6] Dube, E., et al., Winter cover crops and fertiliser effects on the weed seed bank in a low-input maize-based conservation agriculture system. *South African Journal of Plant and Soil,* 2012. 29(3-4): p. 195-197.

[7] Creamer, N.G. and K.R. Baldwin, An evaluation of summer cover crops for use in vegetable production systems in North Carolina. *Hortscience,* 2000. 35(4): p. 600-603.

[8] Falquet, B., et al., Weed Suppression by Common Buckwheat: A Review. *Environmental Control in Biology,* 2015. 53(1): p. 1-6.

[9] Iqbal, Z., et al., Allelopathic activity of buckwheat: isolation and characterization of phenolics. *Weed Science,* 2003. 51(5): p. 657-662.

[10] Kalinova, J. Influence of Common Buckwheat on Growth of Other Plant Species. *in 9th International Symposium on Buckwheat.* 2004. Prague.

[11] Tominaga, T. and T. Uezu, Weed Suppresion by Buckwheat. *Current Advances in Buckwheat Research* 1995: p. 693-697.

[12] Tschuy, F., et al., Unkrautunterdrückung durch Zwischenfrüchte: Analyse verschiedener Faktoren. *Agrarforschung Schweiz,* 2014. 5(7-8): p. 292-299.

[13] Kumar, V., D.C. Brainard, and R.R. Bellinder, Suppression of powell amaranth (Amaranthus powellii), shepherd's-purse (Capsella bursa-pastoris), and corn chamomile (Anthemis arvensis) by buckwheat residues: Role of nitrogen and fungal pathogens. *Weed Science,* 2008. 56(2): p. 271-280.

[14] Kumar, V., D.C. Brainard, and R.R. Bellinder, Effects of Spring-sown Cover Crops on Establishment and Growth of Hairy Galinsoga (Galinsoga ciliata) and Four Vegetable Crops. *Hortscience,* 2009. 44(3): p. 730-736.

[15] Kumar, V., et al., Buckwheat Residue Effects on Emergence and Growth of Weeds in Winter-Wheat (Triticum aestivum) Cropping Systems. *Weed Science,* 2011. 59(4): p. 567-573.

[16] Tsuzuki, E. and Y.J. Dong, Buckwheat allelopathy: Use in weed management. *Allelopathy Journal,* 2003. 12(1): p. 1-11.

[17] Xuan, T.D. and E. Tsuzuki, Allelopathic plants: Buckwheat (Fagopyrum spp.). *Allelopathy Journal,* 2004. 13(2): p. 137-148.

[18] Hayashi, H. Effects of Leaf and Root Exudates of Common Buckwheat on the Early Growth of Weeds. in *VII International Symposium on Buckwheat Advances in Buckwheat Research 1998.* Winnipeg, Manitoba, Canada

[19] Kalinova, J., J. Triska, and N. Vrchotova, Biological activity of phenolic compounds present in buckwheat plants. *Allelopathy Journal,* 2005. 16(1): p. 123-129.

[20] Khanh, T.D., et al., The exploitation of crop allelopathy in sustainable agricultural production. *Journal of Agronomy and Crop Science,* 2005. 191(3): p. 172-184.

[21] Ohsawa, R. and C. Nakatani, Inter- and intra-cultivar variations in the allelopathic effect of leaf aqueous extract of buckwheat (Fagopyrum esculentum Moench) on the growth of lettuce seedling. *Fagopyrum,* 2005. 22: p. 21-24.

[22] Golisz, A., et al., Specific and total activities of the allelochemicals identified in buckwheat. *Weed Biology and Management,* 2007. 7(3): p. 164-171.

[23] Kalinova, J., N. Vrchotova, and J. Triska, Exudation of allelopathic substances in buckwheat (Fagopyrum esculentum Moench). *Journal of Agricultural and Food Chemistry,* 2007. 55(16): p. 6453-6459.

[24] Kato-Noguchi, H., H. Sugimoto, and M. Yamada, Buckwheat Seedlings May Inhibit Other Plant Growth by Allelopathic Substances. *Environmental Control in Biology,* 2007. 45(1): p. 27-32.

[25] Tin, W.W.T., et al., Caprolactam, an Inhibitory Allelochemical Exuded from Germinating Buckwheat (Fagopyrum Esculentum) Seeds. *Heterocycles,* 2009. 78(5): p. 1217-1222.

[26] Ciarka, D., et al., Allelopathic potential of sunflower. II. Allelopathic activity of plants compounds released in environment. *Allelopathy Journal,* 2009. 23(2): p. 243-254.

[27] Rimando, A.M. and S.O. Duke, *Studies on rice allelochemicals., in Rice: Origin, History, Technology and Production.,* C.W. Smith, Dilday, R. H., Editor. 2003, John Wiley & Sons, Inc., Hoboken, New Jersey. p. 221-244.

[28] Sadaqa, E.A., A.A. Bawazir, and J.R. Qasem, Allelopathic activity of some common weeds species in onion fields. *Allelopathy Journal,* 2010. 26(2): p. 175-184.

[29] Jankowski, K., et al., The initial growth and development of poa pratensis under the allelopathic influence of taraxacum officinale. *Journal of Ecological Engineering,* 2014. 15(2): p. 93-99.

[30] Otusanya, O.O., A.A. Sokan-Adeaga, and O.J. Ilori, Allelopathic effect of the root exudates of Tithonia diversifolia on the germination, growth and chlorophyll accumulation of Amaranthus dubius L. and Solanum melongena L. *Research Journal of Botany,* 2014. 9(2): p. 13-23.

[31] Asaduzzaman, M., et al., Canola cultivar performance in weed-infested field plots confirms allelopathy ranking from in vitro testing. *Biocontrol Science and Technology,* 2014. 24(12): p. 1394-1411.

[32] Inderjit, Soil microorganisms: An important determinant of allelopathic activity. *Plant and Soil,* 2005. 274: p. 227-236.

[33] Li, X.G., et al., Soil sickness of peanuts is attributable to modifications in soil microbes induced by peanut root exudates rather than to direct allelopathy. *Soil Biology and Biochemistry,* 2014. 78: p. 149-159.

[34] Bicksler, A.J. and J.B. Masiunas, Canada Thistle (Cirsium arvense) Suppression with Buckwheat or Sudangrass Cover Crops and Mowing. *Weed Technology*, 2009. 23(4): p. 556-563.

[35] Zhang, S.Z., et al., Interference of allelopathic wheat with different weeds. *Pest Management Science*, 2015.

[36] Wang, P., et al., Shoot competition, root competition and reproductive allocation in Chenopodium acuminatum. *Journal of Ecology,* 2014.

[37] Kumar, V., D.C. Brainard, and R.R. Bellinder, Suppression of Powell Amaranth (Amaranthus powellii) by Buckwheat Residues: Role of Allelopathy. *Weed Science,* 2009. 57(1): p. 66-73.

[38] Asaduzzaman, M., et al., Metabolomics differentiation of canola genotypes: Toward an understanding of canola allelochemicals. *Frontiers in Plant Science,* 2015. 5(JAN).

[39] Hiradate, S., et al., Quantitative evaluation of allelopathic potentials in soils: Total activity approach. *Weed Science,* 2010. 58(3): p. 258-264.

[40] Duke, S.O., Allelopathy: Current status of research and future of the discipline: A commentary. *Allelopathy Journal,* 2010. 25(1): p. 17-30.

[41] Duke, S.O., Proving allelopathy in crop-weed interactions. *Weed Science,* 2015. 63: p. 121-132.

[42] Whitehead, D.C., H. Dibb, and R.D. Hartley, Bound phenolic compounds in water extracts of soils, plant roots and leaf litter. *Soil Biology and Biochemistry,* 1983. 15(2): p. 133-136.

[43] Kong, C., et al., Two compounds from allelopathic rice accession and their inhibitory activity on weeds and fungal pathogens. *Phytochemistry,* 2004. 65(8): p. 1123-1128.

In: New Developments in Allelopathy Research
Editor: Julia E. Price

ISBN: 978-1-63483-390-5

Chapter 3

OVERCOMING THE AUTOTOXIC EFFECTS IN HYDROPONIC STRAWBERRY BY AMINO ACIDS SUPPLEMENTATION ON LEAVES AND ELECTRO-DEGRADATION OF NUTRIENT SOLUTION

Md. Fuad Mondal*[1,2,3]*, Md. Asaduzzaman*[1]*,
***Md. Hafizur Rahman Hafiz*[1,4] *and Toshiki Asao*[1,*]**

[1]Department of Agriculture, Faculty of Life and Environmental Science, Shimane University, 2059 Kamihonjo, Matsue, Shimane, Japan
[2]United Graduate School of Agricultural Sciences, Tottori University, Koyama-cho, Minami Tottori, Tottori, Japan
[3]Faculty of Agriculture, Sylhet Agricultural University, Sylhet, Bangladesh
[4]Hajee Mohammad Danesh Science and Technology University, Dinajpur, Bangladesh

ABSTRACT

Autotoxicity is a common problem in strawberry. During this phenomenon strawberry plant's roots secreted allelochemicals to the culture solution causing damage to the root cells, which in turns hamper water and mineral nutrient absorption. As a result, the growth of shoot and root, number of flowers and harvested fruit per plant, and fruit enlargement reduced greatly. In our two studies we have tried to recover the growth and yield of strawberry by the foliar supplementation of amino acids and electro-degradation (ED) of recycle nutrient solution. In both studies, we found the recovery of growth and yield of strawberry during autotoxicity. In our first study, twenty two amino acids were short listed to Ala, Glu, Hyp, Thr, His and Phe on basis of their performance on growth and yield of strawberry in non-recycled hydroponics. Then short listed amino acids and GABA were further investigated following Wagner's pot hydroponics. Among the seven amino acids applied, only Glu and Hyp improved their fruit yield over 50%. The *in vitro* culture of strawberry plantlets showed similar results of previous experiments. Therefore,

* Corresponding author. Tel: +81 852 34 1817; Fax: +81 852 34 1823. *E-mail address:* asao@life.shimane-u.ac.jp (T. Asao).

considering the effects of amino acids on growth and fruit yield of strawberry, Glu and Hyp can be used for foliar application in a non-recycled hydroponic culture. In our second study, we found complete recovery in fruit yield of strawberry plants grown in closed hydroponic system where culture solutions were allowed to ED treatment before reuse. In the growth chamber bioassay strawberry plantlets showed retarded growth in non-renewed and/or its weekly ED whereas it was recovered in biweekly applications. Therefore, ED nutrient solutions were investigated following without plant experiments. It was found that ED of nutrient solution was associated with degradation of Fe-EDTA (about 10%) and low concentration of Ca^{2+} (about 66% of the initial) in the treated culture solution. The disadvantages of longer duration ED were decrease in solution pH as low as 3.1 and increase in temperature. Therefore, shorter duration (2h) ED treatments were applied in the greenhouse experiments following Wagner's pot hydroponics. ED machine used for previous experiments was modified for allowing the treated solution passed continuously through the electrode. Results obtained from Wagner's pot experiment showed that growth and yield (about 99% to renewed control) of strawberry plants were improved in non-renewed nutrient solution having ED at four weeks interval. Therefore, in practice, application of ED to non-renewed nutrient solution in closed hydroponic system for 2 h at every four weeks intervals would recover autotoxicity in strawberry.

1. INTRODUCTON

Autotoxicity is a phenomenon of intraspecific allelopathy that occur when a plant species releases chemical substances that inhibit or delay germination and growth of the same plant species (Putnam 1985a; Miller 1996; Singh et al. 1999). It has been reported to occur in a number of crop plants in agro ecosystem causing serious problems such as growth reduction, yield decline and replant failures (Singh et al. 1999; Pramanik et al. 2000; Asao et al. 2003). Plants when experiences autotoxicity it releases chemicals/ allelochemicals to its rhizosphere (Singh et al. 1999) through various mechanisms. These include leachation (Overland, 1966), volatilization (Petrova, 1977), root exudation (Tang and Young, 1982), crop residue decomposition (Rice 1984; Putnam 1985b), and pollen spreading in some plants (Cruz-Ortega et al. 1988). The released chemical compounds create problems in monoculture and/or closed hydroponic culture systems as they can accumulate and inhibit the growth of the actual crop. Exposure of these allelochemicals play a multitude of ecological and physiological roles as they inhibit plant growth (Rice, 1984), alter mineral uptake (Lyu and Blum, 1990; Baziramakenga et al. 1994), disrupt membrane permeability (Baziramakenga et al. 1995), cause stomatal closure and induce water stress (Barkosky and Einhelling, 1993), influence respiration (Penuelas et al. 1996), affect photosynthesis and protein synthesis (Mersie and Singh 1993; Rohn et al. 2002), impair hormone balance (Holappa and Blum, 1991) and alter enzyme activities (Rohn et al. 2002; Doblinski et al. 2003). Among them, ion uptake and hydraulic conductivity (i.e., water uptake) are worse affected processes since root is the first organ to come into contact with autotoxins in the rhizosphere (Blum et al. 1999).

Closed hydroponic system has recently gained popularity for commercial production of strawberry (Takeuchi 2000; Oka 2002; Koshikawa and Yasuda, 2003) but this managed culture technique has the problem of autotoxicity due to accumulation of allelochemicals in the culture solution. In strawberry autotoxicity from root exudates has been investigated in closed hydroponic culture and benzoic acid was confirmed as the most potent growth inhibitor (Kitazawa et al. 2005). Strawberry plants when experiences autotoxicity, the shoot and root growth, number of flowers and harvested fruits per plant, and fruit enlargement reduces significantly. Elimination of growth inhibitors from recycling hydroponic cultivation system is desirable from the viewpoint of conservation oriented agriculture. Activated charcoal (AC) supplementation was found to improve growth and yield significantly by adsorbing phytotoxic chemicals (mainly phenolics), for example in strawberry (Kitazawa et al. 2005), taro (Asao et al. 2003), cucumber (Yu and Matsui 1994; Asao et al. 1998, 1999, 2000), several leafy vegetables (Asao et al. 2004a), and some ornamentals (Asao et al. 2007). Several other methods has also been found to be effective in removing/degradating the phytotoxic substances released from plant roots during autotoxicity such as degradation by microbial strain (Asao et al. 2004b) or auxin (2,4-D and NAA) supplementation (Kitazawa et al. 2007). However, finding suitable method for controlling autotoxicity in strawberry would be of great help for the commercial production of strawberry in a non-recycled hydroponics. Therefore we have studied two methods of growth and yield recovering in strawberry during the autotxicity in closed hydroponic system in Experimental Research Center at Biological Resources Science, Shimane University. In this chapter we illustrate the growth and yield recovery in strawberry plants under autotoxicity by suppimentation of amino acids on the leaves and through electrodegradation of non renewed nutrient solution.

2. Materials and Methods

2.1. Plant Materials

The virus free and healthy strawberry (*Fragaria* × *ananassa* Duch. var. 'Toyonoka') plantlets obtained through plant tissue culture were used for all experiments.

2.2. Plant Cultivation in Hydroponics

For bioassay, plantlets were planted to plastic containers (17 × 29 × 9.5 cm) in the growth chamber at 20/15°C (day/night) with a light intensity of 74–81 μmol m^{-2} s^{-1} and a 14-hours photoperiod. The containers were filled with 3 L of 25% standard 'Enshi' solutions (pH 7.25 and EC 0.8 dS m^{-1}). The full strength Enshi nutrient solution contains the following amount of salts per 1000 L of tap water: 950 g $Ca(NO_3)_2{\cdot}4H_2O$; 810 g of KNO_3; 500 g of $MgSO_4{\cdot}7H_2O$; 155 g of $NH_4H_2PO_4$; 3 g of H_3BO_3; 2 g of $ZnSO_4{\cdot}7H_2O$; 2 g of $MnSO_4{\cdot}4H_2O$; 0.05 g of $CuSO_4{\cdot}5H_2O$; 0.0 2g of Na_2MoO_4; 25 g of NaFe-EDTA (Hori, 1966). For the growth and recovery experiments plantlets were transplanted either to plastic container (20 × 54 × 34 cm) with 55 liters of 25% Enshi nutrient solution or Wagner's pot hydroponic system. Wagner's pot connected with a plastic reservoir (63 × 48 × 22 cm) containing 60 liters of 25% Enshi nutrient solution in closed hydroponic system. The system includes main inlet pipes (15 mm diameter) for supply and drainage of nutrient solution between reservoir and pots, Wagner's pot (1/5000a, NF-5, AsOne, Osaka, Japan) with 3 liters capacity for planting, inlet tubes (4 mm diameter) to supply solution to the pots, and 60 liters capacity nutrient solution container with a pump (KP-101, Koshin, Kyoto, Japan). The culture solution was recycled through the pipes for 5 minutes at 10 minutes intervals using an automatic pump timer (KS-1500, Iuchi, Osaka, Japan). During the culture period pollination was aided by a soft brush at two days intervals. Fruits were harvested when those became about 80% red in colour. At each harvest fresh weight of fruits were recorded and gathered for final yield calculation. At final harvest, growth parameters were recorded. Then strawberry plant parts were separated into leaf, crown, inflorescence and root and dried in a constant temperature oven (DKN 812, Yamato Scientific Co., Ltd. Japan) for 72 h at 80°C. When the dry matter reaches constant weight, dry weight of different plant parts was measured.

2.3. Determination of Fruit Qualities of Strawberry

Strawberry fruits were composited after each harvest and were frozen at –30°C for subsequent analysis of soluble solids, titratable acids and ascorbic acid content. Fruit samples were kept out of freezer before analysis to obtain juice for determining the above qualities of strawberry fruits. The soluble solid content of the fruit was determined using a digital refractometer (PR-1, Atago Ltd., Japan). Titratable acid contents were determined by diluting each 2 ml aliquot of strawberry juice to 10 ml with 8 ml distilled water and added 2–3 drops of phenolphthalein then adjusted the pH to 8.2 using 0.1 N (w/v) NaOH. The quantity of NaOH (ml), and the amount for appropriate acidity was converted into citric acidity (%). Ascorbic acid content was measured with 2,4-dinitrophenylhydrazine (DNP) colorimetry. Strawberry fruit juice (0.5 mL) were taken in 50 mL test tube then 0.5 mL of 10% meta-phosphoric acid solution, 1 mL of distilled water, 1 mL of 0.03% 2,6-dichlorophenol-indophenol (DCP), 2 mL of thiourea, and 1 mL of 2,4-DNP was added to the samples following 3 h incubation at 37°C in water bath (BW400, Yamato Scientific Co. Ltd. Japan). After incubation samples 5 mL of 85% H_2SO_4 were added keeping in water cooled with iced water. After 30 min cooling ascorbic acid content was measured at 520 nm by spectrophotometer (U-2900, Hitachi High Technologies Corporation, Tokyo, Japan).

2.4. Statistical Analysis

A randomized complete block design with three replicates was used for both culture of strawberry in container based hydroponics and Wagner's pot hydroponics in the greenhouse whereas, complete block design was performed in culture of strawberry plantlets *in vitro* condition. Analysis of variance was performed to test for statistical differences among the treatments and mean separations were performed by Tukey's Honestly Significant Difference (HSD) and Least Significant Difference (LSD) test at $P < 0.05$ level of significance.

3. Recovery from Autotoxicity in Strawberry by Supplementation of Amino Acids

Autotoxicity from the root exudates of strawberry in closed hydroponic culture has been investigated (Kitazawa et al., 2005). During this phenomenon strawberry plant's roots secreted allelochemicals mainly benzoic acid to the culture solution causing damage to the root cells, which in turns hamper water and mineral nutrient absorption. Allelopathic compounds may induce a secondary oxidative stress manifested as enlarged production of reactive oxygen species (ROS) (Weir et al., 2004). Toxic ROS can affect membrane permeability, cause damage to DNA and protein, induce lipid peroxidation, and ultimately lead to programmed cell death. Therefore, autotoxic effects of root exudates of strawberry plants on its growth and development is likely to be caused by impairment of nutrient and water absorption by injured roots. Supply of mineral nutrient alternatively other than by root uptake can sustain plant growth during this allelochemical stress. The availability and uptake of nitrogen is considered as the major factor affecting growth (Lea and Azevedo, 2006) therefore, it can be sprayed on the leaves as a source of nutrients. Use of foliarly applied urea as a nitrogen source is common (Bowman and Paul, 1992; Vasilas et al., 1980). For example, in wheat, foliarly applied urea produced positive effects; these were attributed to higher leaf photosynthetic rates and higher leaf urease enzyme activities (Peltonen, 1993).

Amino acids are the nitrogenous compound which forms the basic component of all living cells. It can absorb by leaf exogenously (Furuya and Umemiya, 2002). Therefore, it has a great potentiality of using under managed culture techniques. Recently they are used as foliar spray to improve the growth, yield and quality of crops (Mazher et al., 2011; Takeuchi et al., 2008). Several researchers found positive impacts of amino acids as foliar spray under stress condition for example Proline to wheat (Rajagopal and Sinha, 1980), Proline, Alanine, Serine, and Asparagine to maize (Thakur and Rai, 1985) under osmotic stress and Proline, Phenylalanine to maize and board bean under salinity stress (Abd El-Samad et al., 2011). As the accumulated allelochemicals in closed culture become stressful to plant, spraying of amino acid to strawberry plants would be positive. So far, spraying amino acids in recovering strawberry plant growth during autotoxicity has not been studied. Therefore, the purpose of the present study was to evaluate the performance of amino acids on the recovery of growth and yield of strawberry plants under autotoxicity in closed hydroponic culture.

3.1. Evaluation of Twenty Two Amino Acids on the Growth of Strawberry Plants under Autotoxicity in Container Based Hydroponics

The study was conducted in 100 m^2 glasshouse of Experimental Research Center at Biological Resources Science, Shimane University. Initially strawberry plantlets at four to five leaves stage were transplanted to plastic container (20 × 54 × 34 cm) with 55 L of 25% Enshi nutrient solution. Five plantlets were planted in each container in such a way that the roots were inserted into the nutrient solution inside the container keeping shoot outside. Urethane foam block (23 mm × 23 mm × 27 mm) was used for holding the plant tight with a floating board on the nutrient solution. Nutrient solutions were circulated 24 h by pumps (KP-101, Koshin, Kyoto, Japan) with automatic timer (KS-1500, Iuchi, Osaka, Japan) which were either renewed or non-renewed entirely and non-renewed with amino acids and urea application. Renewed culture solutions were changed biweekly with new nutrient solutions whereas, non-renewed nutrient solutions were analyzed for major nutrients and adjusted as close as possible to initial concentrations at every two weeks on the basis of chemical analyses with Compact NO_3^- meter (B-343, Horiba, Ltd. Kyoto, Japan) for NO_3^-, Spectrophotometer (U-2900, Hitachi, Tokyo, Japan) for PO_4^{3-} and Polarized Zeeman Atomic

Absorption Spectrophotometer (Z-2310, Hitachi, Tokyo, Japan) for K^+, Ca^{2+}, Mg^{2+} and Fe^{3+}. Two day after transplanting, twenty two water soluble amino acids viz., Alanine (Ala), Arginine (Arg), Asparagine (Asn), Aspartic acid (Asp), Cysteine (Cys), Glutamatic acid (Glu), Glutamine (Gln), Glycine (Gly), Hydroxy-proline (Hyp), Lysine (Lys), Ornithine (Orn), Proline (Pro), Serine (Ser), Threonine (Thr), Tryptophan (Typ), Methionine (Met), Leucine (Leu), Isoleucine (Ile), Citrulline (Cit), Histidine (His), Phenylalanine (Phe), and Valine (Val) and urea were individually sprayed as foliar application at 2 mL per plant by a 500 mL sprayer three times a week on the strawberry plants grown in non-renewed nutrient solution. The concentrations of urea and amino acids were adjusted to nitrogen content of Pro at 200 mg L^{-1} to maintain the same concentration level. The dates of anthesis were recorded for each plant to check whether any influence of amino acids on flowering of strawberry among the treatments.

Fruits were harvested when those became about 80% red in colour. At each harvest fresh weight of fruits were recorded and gathered for final yield calculation. At final harvest, leaf number, leaf length and width, root length, crown diameter, fresh weight of leaf, crown and inflorescence were recorded. Then strawberry plant parts were separated into leaf, crown, inflorescence and root and dried in a constant temperature oven (DKN 812, Yamato Scientific Co., Ltd. Japan) for 72 h at 80°C. When the dry matter reaches constant weight, dry weight of different plant parts was measured.

Foliar application of twenty two amino acids showed significant influence on the growth of the strawberry plants grown in non-renewed culture solution in hydroponics (Table 1). Plants grown in renewed nutrient solution produced bigger leaves compared to plants grown in non-renewed nutrient solution. Leaf size was not significantly increased by foliar spray with Arg, Asp, Gln, Gly, Orn, Pro, Ser, Met, Leu, Ile, Cit and Val. Spray of Ala, Cys, Glu, Hyp, Lys, Thr, Trp, His and Phe on the leaves of strawberry plants grown in non-renewed nutrient solution increased leaf length and width compared with water as control. Number of leaves did not differ significantly in plants grown in non-renewed nutrient solution with or without supplementation of amino acids. Longer roots were recorded in plants sprayed with urea, Ala, Cys, Lys, Trp and grown in renewed nutrient solution compared to other amino acids.

Smaller crowns were found in Gly, Pro, Ser, Met, Leu and Val sprayed plants compared to plants grown in renewed culture solution. Higher leaf fresh weight was measured in Cys, Glu, Trp and Phe treated plants compared to plants grown in non-renewed solution. Crown fresh weight was increased in plants sprayed with urea, Ala, Cys, Glu, Gln, Hyp, Lys, Thr, Trp, Cit, His, Phe and Val whereas, fresh weight of flowering bud was increased in plants sprayed with Cys, Glu, Hyp, Lys, Thr, Trp, His and Phe. Among the amino acids applied Cys, Glu, Lys, Trp, His and Phe supplemented plants produced higher dry weight in leaves, crown, inflorescence and root compared to plants in non-renewed nutrient without amino acid application (Figure 1). Ala, Hyp and Thr also produced higher dry matter in all parts except leaves.

From above results it is evident that growth variables were reduced when strawberry plants were grown in non-renewed nutrient solutions compared to renewed solution but these were improved in plants grown in non-renewed nutrient solution with the supplementation of Ala, Cys, Glu, Hyp, Lys, Thr, Trp, His and Phe.

Table 1. Effect of twenty two amino acids on the growth of strawberry plants grown in non-renewed nutrient solution in closed hydroponic system

Amino acids[a]	Number of leaves[b]	Leaf length (cm)	Leaf width (cm)	Root length (mm)	Crown diameter (mm)	FW of leaves (g)	FW of crown (g)	FW of flowering bud (g)
RW	42.4 ab[c]	37.4 a	21.7 a	372.5 ab	49.5 a	194.6 a	11.7 a	14.6 a
NRW	24.0 ab	17.0 c	13.3 bc	283.7 bc	39.8 ab	26.3 c	4.5 b	5.1 bc
Urea	28.1 ab	27.0 b	18.7 ab	401.5 a	39.5 ab	73.9 bc	8.7 ab	6.6 bc
Ala	38.4 ab	29.5 ab	20.2 ab	334.0 ab	36.5 ab	121.2 bc	9.1 ab	8.9 b
Arg	33.0 ab	17.9 c	12.9 bc	234.5 c	38.0 ab	36.0 c	5.4 b	6.4 bc
Asn	30.0 ab	26.3 b	17.9 ab	297.5 bc	35.5 ab	56.5 c	4.5 b	6.7 bc
Asp	33.6 ab	25.4 bc	18.1 ab	281.9 bc	32.1 ab	73.9 bc	6.1 b	8.2 bc
Cys	40.2 ab	33.3 ab	20.3 ab	335.0 ab	41.2 ab	147.2 ab	9.5 ab	10.9 ab
Glu	37.8 ab	33.2 ab	21.5 ab	314.0 b	38.5 ab	146.7 ab	9.2 ab	12.8 ab
Gln	25.3 ab	25.5 bc	17.0 b	320.5 b	34.6 ab	56.7 c	6.9 ab	6.3 bc
Gly	23.4 ab	15.8 c	12.5 bc	282.3 bc	29.7 b	28.2 c	5.8 b	3.3 bc
Hyp	36.0 ab	30.6 ab	19.5 ab	317.5 b	41.5 ab	105.7 bc	8.7 ab	11.1 ab
Lys	34.8 ab	33.3 ab	21.6 ab	336.0 ab	43.5 ab	128.3 b	9.0 ab	11.5 ab
Orn	24.5 ab	23.7 bc	15.4 bc	306.5 b	39.8 ab	40.1 c	5.4 b	4.8 bc
Pro	23.6 b	20.1 bc	15.3 bc	300.5 bc	23.9 b	39.5 c	4.3 b	4.0 bc
Ser	25.8 ab	19.8 bc	16.3 bc	292.1 bc	23.5 b	48.6 c	5.6 b	5.9 bc
Thr	34.9 ab	32.0 ab	20.5 ab	304.5 b	43.5 ab	100.8 bc	8.9 ab	9.9 ab
Trp	43.7 a	35.1 ab	21.4 ab	374.5 ab	47.0 ab	139.5 ab	11.4 ab	10.9 ab
Met	28.2 ab	19.0 bc	13.7 bc	293.0 bc	28.6 b	32.5 c	4.3 b	3.5 bc
Leu	31.2 ab	10.1 c	11.8 c	276.3 bc	27.5 b	29.4 c	3.6 b	2.6 c
Ile	32.9 b	21.2 bc	14.1 bc	297.0 bc	36.0 ab	49.5 c	5.0 b	5.7 bc
Cit	22.7 ab	15.8 c	13.7 bc	283.5 bc	33.0 ab	23.1 c	6.6 ab	3.9 bc
His	39.5 ab	32.3 ab	21.0 ab	319.0 b	36.6 ab	123.4 b	11.1 ab	10.4 ab
Phe	41.4 ab	33.4 ab	21.8 a	329.0 b	39.3 ab	145.3 ab	9.2 ab	11.4 ab
Val	27.4 ab	20.1 bc	14.6 bc	279.5 bc	28.8 b	37.3 c	7.8 ab	4.8 bc

[a]Strawberry plants grown in renewed (RW), non-renewed (NRW), and non-renewed nutrient solution with amino acids and urea application. [b] Parameters were measured on per plant basis; Fresh weight (FW), Dry weight (DW). [c] Means within column followed by different letters are significant according to Tukey's test at $P < 0.05$ ($n = 15$). * Significant at $P < 0.05$.

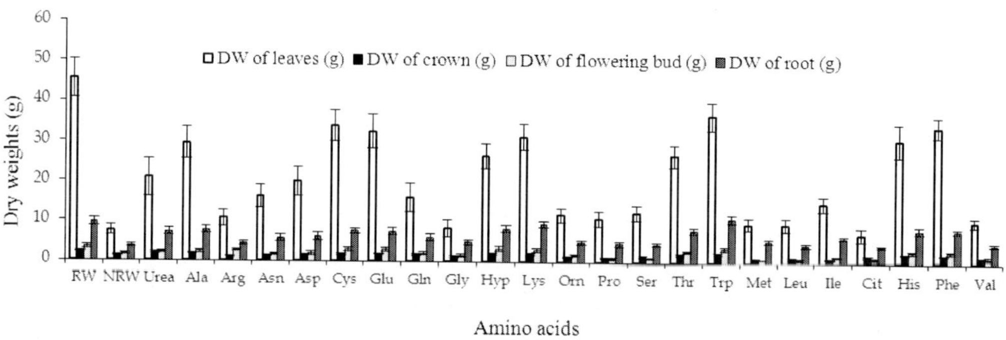

Figure 1. Effects of twenty two amino acids on the dry matter production of strawberry plants grown in non-renewed nutrient solution in closed hydroponic system. RW: Renewed, NRW: Non-renewed, amino acids are presented as their three letters abbreviation.

Among the amino acids applied Cys, Glu, Lys, Trp, His and Phe supplemented plants produced higher dry weight in leaves, crown, inflorescence and root compared to plants in non-renewed nutrient without amino acid application (Figure 1). Ala, Hyp and Thr also produced higher dry matter in all parts except leaves. From above results it is evident that growth variables were reduced when strawberry plants were grown in non-renewed nutrient solutions compared to renewed solution but these were improved in plants grown in non-renewed nutrient solution with the supplementation of Ala, Cys, Glu, Hyp, Lys, Thr, Trp, His and Phe.

Table 2. Effect of twenty two amino acids on fruit quality of strawberry plants grown in non-renewed nutrient solution in closed hydroponic system

Amino acids[a]	Soluble solids (%)	Citric acidity (%)	Ascorbic acid (mg/100g)
RW	7.5	0.42	61.3
NRW	5.8	0.38	69.3
Urea	7.2	0.45	55.9
Ala	7.6	0.45	57.3
Arg	8.3	0.45	54.2
Asn	7.8	0.48	55.1
Asp	6.5	0.42	54.5
Cys	6.7	0.38	74.1
Glu	7.2	0.38	60.8
Gln	7.2	0.42	64.1
Gly	7.0	0.35	57.5
Hyp	6.9	0.35	55.4
Lys	7.1	0.26	55.0
Orn	6.0	0.38	58.3
Pro	6.4	0.32	50.1
Ser	7.8	0.35	44.5
Thr	7.5	0.32	50.8
Trp	8.0	0.35	58.4
Met	7.1	0.42	76.3
Leu	7.8	0.42	64.6
Ile	8.0	0.54	74.9
Cit	7.3	0.48	60.9
His	7.4	0.35	78.6
Phe	7.5	0.42	69.3
Val	7.4	0.35	72.0
	Ns	ns	ns

[a] Strawberry plants grown in renewed (RW), non-renewed (NRW), and non-renewed nutrient solution with amino acids and urea application. [b] Parameters were measured on per plant basis. [ns, *] Non-significant according to Tukey's test at $P < 0.05$ (n = 1).

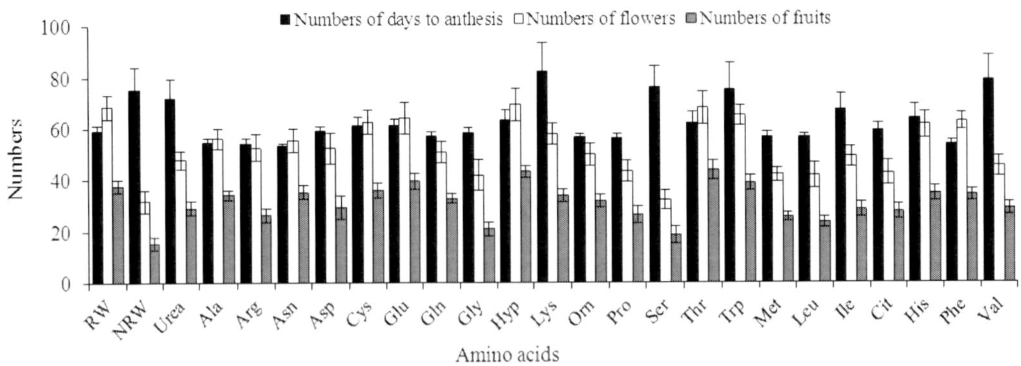

Figure 2. Effect of twenty two amino acids on the flowering and fruit setting of strawberry plants grown in non-renewed nutrient solution in closed hydroponic system. RW: Renewed, NRW: Non-renewed, amino acids are presented as their three letters abbreviation.

Amino acids lead a positive effect on the yield attributes and fruit quality of strawberry grown in container based hydroponics (Table 2; Figure 2, 3). Anthesis date was influenced by the application of amino acids on strawberry plant leaves and it was found that about 23 days earlier flowering in Ala, Arg, Asn and Phe sprayed plants. Fruit yield per plant was decreased about 74% in plants grown in non-renewed nutrient solution than grown in renewed nutrient solution. Application of urea, Ala, Asn, Asp, Cys, Glu, Gln, Hyp, Lys, Orn, Thr, Trp, His, Phe and Val improved fruit yield in plants grown in non-renewed nutrient solution which was attributed by number of flowers and number of mature fruits. Average fruit weight also correspond the yield in these amino acid supplemented plants. There were no significant differences among the amino acid applied in terms of strawberry fruit qualities such as soluble solids, citric acidity and ascorbic acid.

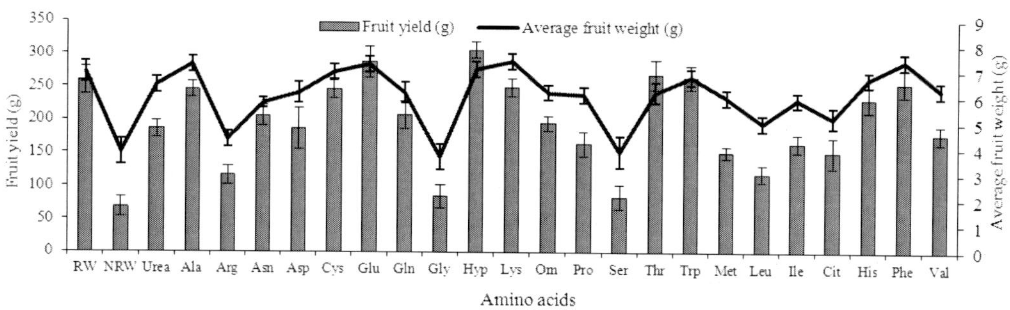

Figure 3. Effect of twenty two amino acids on the yield of strawberry plants grown in non-renewed nutrient solution in closed hydroponic system. RW: Renewed, NRW: Non-renewed, amino acids are presented as their three letters abbreviation.

3.2. Effects of Selected Seven Amino Acids on the Growth of Strawberry Plants under Autotoxicity in Wagner's Pot Hydroponics

Seven amino acids were selected for their better growth and yield performance in the container based strawberry hydroponics. These short listed amino acids were further investigated in the glasshouse following Wagner's pot hydroponics. Healthy plantlets with five to six leaves were planted into Wagner's pots. The culture solution was not renewed during the entire growth period and it was recycled. One month after transplanting, the selected amino acids viz., Ala, Glu, Hyp, Thr, His, Phe, including *gamma*-aminobutyric acid (GABA) and water as control were sprayed on leaves of strawberry plants. The concentrations of all amino acids were adjusted to nitrogen content of Pro at 200 mg L^{-1}. Water and amino acids were applied at 1.4 mL per plant by a 100 mL sprayer three times a week. The major nutrients in the non-renewed culture solution were analyzed and adjusted following methods and instruments used in container based hydroponics at every two weeks. Other cultural practices were done as described in the previous culture. Fruits were harvested when those became about 80% red in colour. The harvested fruits were grouped into three stages based on their harvesting date and gathered for final yield calculation.

The relative amount of chlorophyll in strawberry leaves were measured (SPAD-502 plus, Konica Minolta Sensing, Inc. Osaka, Japan) at final harvest. Growth and yield of strawberry plants were measured following the methods as described in the previous culture.

Table 3. Effect of seven amino acids on the growth of strawberry plants grown in non-renewed nutrient solution in Wagner's pot hydroponic system

Amino acids[a]	Number of leaves[b]	SPAD	FW of leaves (g)	FW of crown (g)	Crown diameter (mm)	Root length (mm)	DW of leaves (g)	DW of crown (g)	DW of root (g)
Water	72.6	38.8	223.2 b[c]	38.9	15.9 d	32.6 ab	51.3 b	7.5 b	10.5 c
Ala	91.3	38.7	278.1 a	44.0	17.0 cd	30.2 ab	63.0 ab	8.5 ab	12.0 bc
Glu	84.5	38.4	272.8 a	45.9	22.0 a	32.5 ab	67.9 a	10.2 ab	14.0 ab
Hyp	87.8	39.5	269.6 ab	49.8	21.8 ab	32.0 ab	64.5 ab	11.3 a	16.7 a
Thr	60.2	38.5	256.1 ab	49.0	17.4 cd	28.4 b	63.4 ab	8.8 ab	11.8 bc
His	88.5	38.4	268.1 ab	47.6	16.5 cd	32.0 ab	66.9 a	8.2 b	12.2 bc
Phe	94.1	39.2	279.0 a	48.2	18.3 cd	30.5 ab	64.2 ab	10.2 ab	14.8 ab
GABA	94.8	39.2	274.0 a	49.2	18.9 bc	34.3 a	63.9 ab	9.8 ab	13.5 abc
	ns	ns	*	ns	*	*	*	*	*

[a] Strawberry plants grown in non-renewed nutrient solution with amino acids application.

[b] Parameters were measured on per plant basis; Fresh weight (FW), Dry weight (DW).

[c] Means within column followed by different letters are significant according to LSD test at $P < 0.05$ ($n = 15$).

[ns, *] Non-significant or significant at $P < 0.05$, respectively.

Table 4. Effect of seven amino acids on the fruit yield of strawberry in Wagner's pot hydroponic system

Amino acids[a]	Number of fruits[b]	Average fruit weight (g)	Fruit yield (g)
Water	63.3 c[c]	6.7	432.2 d[z]
Ala	88.2 ab	6.8	560.2 abc
Glu	90.4 ab	7.3	654.2 a
Hyp	97.5 a	6.8	649.8 a
Thr	90.5 ab	6.7	606.2 ab
His	89.2 ab	7.0	594.7 ab
Phe	77.3 bc	6.9	474.1 cd
GABA	72.8 bc	7.3	498.2 bcd
	**	ns	**

[a] Strawberry plants grown in non-renewed nutrient solution with amino acids application.

[b] Parameters were measured on per plant basis; Fresh weight (FW), Dry weight (DW).

[c] Means within column followed by different letters are significant according to LSD test at $P < 0.05$ ($n = 15$).

[ns, **] Non-significant or significant at $P < 0.01$, respectively.

Result showed a significant difference in plants growth supplemented with amino acids (Table 3). Among the amino acids, Ala, Glu, Phe and GABA supplemented plants increased their leaf fresh weight by about 25, 22, 25, and 23%, against water spray as control, respectively whereas leaf dry weight was significantly increased only in Glu and His treated plants. There was no significant difference among the treatments in terms of number of leaves, relative amount of chlorophyll content and crown fresh weight.

Bigger crown was found in Glu (22.0 mm), Hyp (21.8 mm) and GABA (18.9 mm) treated plants. Foliar spray of Hyp significantly increased the crown dry weight. Amino acids have influence on the root growth which was evidenced in Glu, Phe and Hyp supplemented plants where these amino acids increased by 33, 41 and 59% dry weight of root, respectively compared with control.

Table 5. Soluble solid content, % citric acidity and ascorbic acid content in fruits of strawberry plants supplemented with amino acids in Wagner's pot hydroponic system. Fruits harvested at stage I (2011/3/25–2011/4/30), stage II (2011/5/1–2011/5/25) and stage III (2011/5/26–2011/6/20)

Amino acids[a]	Soluble solid content (%)			Citric acidity (%)			Ascorbic acid (mg/100g)		
	Stage I	Stage II	Stage III	Stage I	Stage II	Stage III	Stage I	Stage II	Stage III
Water	7.6	7.3	7.9	0.22 b[b]	0.22 c	0.26 b	42.1 ab	41.0	40.3
Ala	6.7	7.1	7.0	0.35 ab	0.26 bc	0.45 a	45.6 a	39.6	39.3
Glu	7.9	7.3	7.1	0.51 a	0.38 ab	0.35 ab	41.6 ab	38.0	36.2
Hyp	6.8	5.8	7.3	0.45 a	0.35 abc	0.41 a	40.9 bc	38.4	35.7
Thr	6.0	5.6	6.8	0.38 ab	0.32 abc	0.35 ab	37.8 bc	41.2	38.4
His	7.4	7.3	7.2	0.38 ab	0.29 bc	0.22 b	40.9 bc	42.6	38.6
Phe	6.8	6.8	6.7	0.38 ab	0.29 bc	0.35 ab	37.0 c	41.2	37.4
GABA	7.4	6.0	7.8	0.58 a	0.45 a	0.45 a	41.7 ab	38.8	40.7
	ns	ns	ns	**	**	**	*	ns	Ns

[a] Strawberry plants grown in non-renewed nutrient solution with amino acids application.

[b] Means within column followed by different letters are significant according to LSD test at P < 0.05 (n = 15).

[ns, *, **] Non-significant or significant at P < 0.05, 0.01%, respectively.

Application of amino acids greatly influenced the yield in strawberry plants following recycled Wagner's pot hydroponics (Table 4). Ala, His, Glu and Hyp treated plants produced higher fruit yield by about 51, 50, 40, 38 and 30% against control, respectively. Spraying of Phe and GABA did not influenced on the fruit yield of strawberry. Numbers of fruits were also increased in amino acids treated plants. Greater numbers of fruits were recorded from Ala, Thr, His, Glu and Hyp treated plants. Average fruit weight was not significantly improved by the amino acids under investigation. Application of amino acids did not left any effects on the soluble solid content at different stages of harvested strawberry fruits, however, titratable acid content in fruits significantly varied in all three stages (Table 5). In the stage I, Glu, Hyp and GABA treated strawberry plants produced fruits with high acid content, stage II fruits analyses showed higher acid content when the plant were supplemented with Glu and GABA. In stage III, the higher acid levels were found in fruits with, Ala, Hyp and GABA application. Although ascorbic acid content in the strawberry fruits was varied in the early harvested fruits but in the mid and later harvested fruits did not differ it significantly.

3.3. Effects of Seven Amino Acids on the Growth of Strawberry Plantlets under *In Vitro* Condition

In order to control the effects of environmental factors and also microbial degradation of amino acid, strawberry plantlets were cultured under *in vitro* condition at Plant Factory supported Research Laboratory of Shimane University. Strawberry cv. Toyonoka plantlets of similar vigor were transferred into a culture box (100 ×110 ×100 mm) with 100 mL substrate. The plant boxes were capped with bio-filter (for aeration) and placed in growth chamber at 20/15 °C (day/night) under florescent light with intensity of 74-81 µmol m^{-2} $s^{-1}$and 12 h photoperiod. The substrates were prepared using 25% Enshi nutrients solution (EC 0.8 dS m^{-1}) with agar (9 g L^{-1}) as solidified agent and sucrose (30 g L^{-1}) as carbon source. One plantlet was planted in each plant box and three plant boxes were used for each treatment with three replications. The seven amino acids (Ala, Glu, Hyp, Thr, His, Phe and GABA) used in the Wagner's pot hydroponics were also used in this experiment with water, urea and renew of substrate as control. In case of renew of substrate, it was changed to new substrate whereas, water, urea and amino acids were sprayed on the strawberry leaves at 3.0 mL per plant using 0.45 µm syringe filter (Toyo Roshi kaisha, Ltd. Japan) inside the clean bench. The concentrations of urea and amino acids used were adjusted to nitrogen content of Pro at 200 mg L^{-1}. After eight weeks, growth variables and relative amount of chlorophyll content in strawberry plants were measured and compare among the treatments.

Growth of strawberry plantlets was evaluated under *in vitro* condition with the supplementation of seven amino acids (Table 6). Leaf number was increased in amino acids supplied plantlets compared to water control. Higher numbers of leaves were counted in plantlets cultured in renewed substrates as control. Urea, Ala, Glu, Phe and GABA treated plantlets increased their relative amount of chlorophyll content against water control. The longest leaf was found in renewed substrate plantlets and all the amino acids supplied plantlets increased their leaf length over water control. Compared with water and urea, Thr improved the leaf width. Ala, Glu, Hyp and GABA treated plantlets resulted in longer roots against water and urea while the longest root was found in renewed substrate plantlets. The results showed that only Hyp application improved the crown diameter. Treated with Hyp produced bigger crown compared to other amino acids. Higher fresh weight of leaf was obtained when plantlets were sprayed by Glu and Hyp as compared to water spray as control. Hyp treated plantlets increased their crown and root fresh weight against water control. All the amino acids treated plantlets improved their root fresh weights against water control. Hyp treated plantlets also produced significantly higher root fresh weight than plantlets grown in renewed substrates. In case of leaf dry weight, Hyp and urea treated plantlets gained higher weight which are similar. Crown dry weight did not showed any significant difference among the amino acid treatments. Under *in vitro* condition strawberry plantlets improved the root dry matter against water and renewed control. Highest root dry weight was found in Glu treated plant and other amino acids such as Ala, Thr, Hyp and GABA; as well as urea also improved.

When plants experiences autotoxicity, ion uptake and hydraulic conductivity (i.e., water uptake) are worse affected processes since root is the first organ to come into contact with autotoxins in the rhizosphere (Blum et al., 1999). Alternative means of supplying mineral nutrient other than absorption by roots can overcome this problem for sustainable growth and yield of strawberry. Studies on the effects of amino acids on the growth and yield of strawberry plants in closed hydroponics would be interesting. In container based hydroponics when twenty two amino acids were sprayed, dry weight of strawberry plants were increased (Figure 1) which accord with the results of Nassar et al. (2003) and Amin et al. (2011) where foliar application of amino acids increased the dry weight of bean and onion plants respectively. As amino acids are the precursor of chlorophyll synthesis, it plays active role in dry matter production in plants (Yaronskaya et al., 2006).

Table 6. Effect of seven amino acids on the growth of strawberry plantlets under *in vitro* condition

Amino acids[a]	Number of leaves[b]	SPAD	Leaf length (mm)	Leaf width (mm)	Root length (mm)	Crown diameter (mm)	FW of leaves (mg)	FW of crown (mg)	FW of root (mg)	DW of leaves (mg)	DW of crown (mg)	DW of root (mg)
RW[x]	12.6 a[c]	40.7 bc	53.2 a	45.8 a	181.2 a	3.5 b	1480.0 a	154.0 b	962.0 c	284.0 a	28.0	102.0 c
Water	6.9 c	42.7 c	43.2 c	37.8 c	81.4 e	3.4 b	967.8 c	125.6 b	691.1 d	201.1 c	27.8	97.8 c
Urea	9.3 b	48.0 ab	47.6 b	38.0 c	86.0 de	3.8 b	1185.6 bc	114.4 b	1470.0 a	270.0 ab	31.1	164.4 a
Ala	8.7 b	48.6 a	48.9 b	40.8 bc	116.8 bc	3.7 b	1049.2 bc	147.0 b	1126.7 bc	220.0 bc	30.0	153.3 ab
Glu	8.6 bc	48.1 ab	52.3 ab	41.5 bc	130.1 b	3.9 b	1194.7 b	169.7 b	1112.8 bc	228.3 bc	34.4	167.8 a
Hyp	9.6 b	44.7 bc	46.9 b	41.0 bc	114.3 bc	4.7 a	1218.3 b	255.0 a	1304.2 ab	267.5 ab	39.2	150.8 ab
Thr	8.9 b	44.2 bc	47.3 b	42.4 ab	103.3 cd	3.8 b	1048.0 bc	142.9 b	1137.5 bc	220.0 bc	29.2	154.2 ab
His	9.3 b	46.4 bc	43.9 b	37.8 c	91.3 de	4.0 ab	1049.2 bc	140.0 b	1005.8 c	218.6 bc	35.0	124.2 bc
Phe	9.3 b	47.8 ab	45.2 b	38.9 bc	101.1 cde	3.9 ab	1078.9 bc	142.2 b	1060.6 bc	224.8 bc	28.3	123.3 bc
GABA	9.4 b	46.8 ab	47.8 b	40.9 bc	122.2 bc	4.0 ab	1179.2 bc	175.8 ab	1228.3 abc	233.3 abc	32.5	155.8 ab
	*	*	*	*	*	*	*	*	*	*	ns	*

[a] Strawberry plants cultured in renewed (RW) and non-renewed substrate with amino acids and water application.

[b] Parameters were measured on per plant basis; Fresh weight (FW), Dry weight (DW).

[c] Means with column followed by different letters are significant according to Tukey's test at $P < 0.05$ ($n = 9$).

[ns, *] Non-significant or significant at $P < 0.05$, respectively.

Moreover foliar application of amino acid increased plant protein content which ultimately increased the dry matter (Das et al., 2002). The regulatory effects of certain amino acids, like Phe and Orn, on plant development through their influence on gibberellins has been suggested by Waller and Nowacki (1978). Plants grown in non-renewed nutrient solution showed growth and yield declined but when plants were supplied with Hyp, it produced higher growth and fruit yield all three cultures. The possible reason might be its presence in the cell wall as Hyp-rich glycoproteins, is an extra-cellular structural protein of plant cell walls and extra-cellular matrix during normal development and in response to stress, autotoxicity in this case (Kieliszewski, 2001; Kieliszewski and Shpak, 2001). The higher fruit yield in amino acid supplemented plants than water sprayed plants were due to greater vegetative growth. This positive effect on growth and yield might be due to the assimilation and metabolism of nitrogen in strawberry plants. For example, Glu is known to have a central role in nitrogen metabolism and is the preferential amino-donor for the different amino transferase reactions for subsequent amino acid interconversions (Lea and Ireland, 1999). Therefore, greater fruit yield was contributed by vigorous growth, number of flowers, number of mature fruits per plant, and average fruit weight.

In Wagner's pot hydroponics, results revealed that the total dry weight (data not shown) was higher in plants sprayed with Glu, Hyp, and Phe than plants grown in non-renewed nutrient solution with water spray. This accord with Mazher et al. (2011) who reported foliar application of Glu increased the growth and the content of total carbohydrate, nitrogen, and phosphorus and potassium percentages of *Codiaeum variegatum* L. plant. In another study, spraying of Pro or Phe on maize and broad bean increased the amount of dry matter and water content (Abd El-Samad et al., 2011). Higher yield was recorded in Ala, Glu, Hyp, Thr, and His treated plants (Table 4) which were attributed by their better vegetative growth and higher numbers of mature fruits per plant. Moreover, amino acids might have some influence on the pollination and fruit setting. Hyp was found to be localized in growing tips in lily which can elongate the pollen tube enhancing the fertilization and fruit setting (Dashek and Harwood, 1974). *In vitro* culture of strawberry plantlets was conducted to exclude the effects of environmental factors like temperature, light intensity, relative humidity and also microbial degradation of amino acids. Therefore, this experiment under control condition can confirm whether there or not any effects of amino acid on the strawberry plant growth in non-renewed nutrient condition. Total dry matter production was greater in urea, Hyp, Glu and GABA treated plantlets compared to water as control (data not shown) which primarily due to meeting the nitrogenous demand from amino acid source. Recent studies found the positive impact of amino acids *in vitro* condition as organic source of nitrogen in alfalfa, maize, sorghum, pineapple, rice and sugarcane (Skokut et al., 1985; Claparols et al., 1993; Rao et al., 1995; Hamasaki et al., 2005; Grewel et al., 2006 and Asad et al., 2009).

4. Growth and Yield Recovery in Strawberry Plants under Autotoxicity through Electrodegradation

Electrochemical methods have been applied for degradation/ oxidation of phenols and its derivatives from organic waste or pollutants by several researchers. Phenolic compounds including phenols, catecol and hydroquinone in aqueous solution and even benzene were found to decompose when treated by electro-degradation (Fleszar and Ploszynka 1985; Comninellis and Pulgarin 1991; Feng and Li, 2003). These compounds are oxidized rapidly at the anode and decompose to CO2. Thus electrodegradation may cause the decomposition of allelochemicals; including benzoic acid exuded in the nutrient solution from plants, and could be useful tool to mitigate autotoxicity in strawberry. Based on these results, previously ED has been tried to decompose benzoic acid (Asao et al. 2008). It was found that exogenously added benzoic acid to a nutrient solution was almost completely decomposed within 24 h. In the ED process a titanium plate was used as cathode and a ferrite rod as anode. The growth inhibitions of plantlets in the nutrient solution containing benzoic acid or in the nutrient solution used for strawberry culture were significantly ameliorated by ED application. However, complete yield recovery from non-renewed nutrient solution to renewal control was suggested through further investigations. Therefore, the present study aimed to determine the appropriate timing and intensity of ED treatment to maintain nutrient solution and allow vigorous plant growth in a closed hydroponic culture. A modified ED machine was used to degrade allelochemicals from the culture solution of strawberry in Wagner's pot hydroponic system.

4.1. Verification of Electrolysis Condition

Electrolysis conditions were verified in a series of concentrations (0, 6.25, 12.5, 25, 50, 75, and 100%) of 'Enshi' nutrient solution (Table 7).

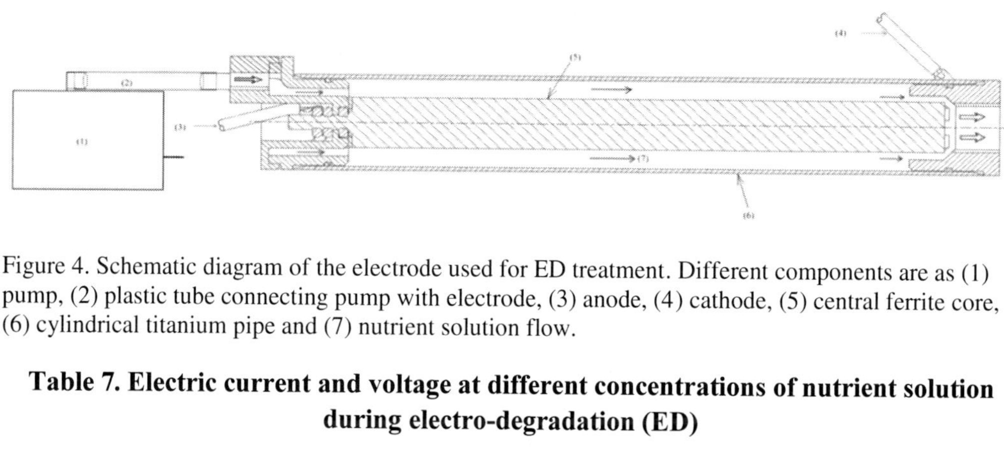

Figure 4. Schematic diagram of the electrode used for ED treatment. Different components are as (1) pump, (2) plastic tube connecting pump with electrode, (3) anode, (4) cathode, (5) central ferrite core, (6) cylindrical titanium pipe and (7) nutrient solution flow.

Table 7. Electric current and voltage at different concentrations of nutrient solution during electro-degradation (ED)

Nutrient solution concentration (%)	Initial		Adjusted	
	Electric current (A)	Voltage (V)	Electric current (A)	Voltage (V)
0.00	1.75*	35.7	2.0	40.2
6.25	1.82	21.0	2.0	22.5
12.50	1.84	16.2	2.0	17.2
25.00	1.86	12.0	2.0	12.6
50.00	1.87	9.8	2.0	10.2
75.00	1.88	8.8	2.0	9.0
100.00	1.88	8.1	2.0	8.4

* Values are the average of three readings.

Initially the electric current was slightly increased with the increase of nutrient solution concentration but voltages were decreased gradually. The electric current ranged from 1.75 to 1.88 A, and electric voltage form 35.7 to 8.1 V. However, the electric current was below 2 A in all the concentrations of nutrient solution. When the electric current was adjusted at 2 A, then the voltage was not changed greatly from the initial. Therefore, 2 A were selected as the test current for the investigations.

4.2. Growth Chamber Bioassay of Strawberry Plantlets

Strawberry plantlets were used for this bioassay. The virus free and healthy plantlets at 3rd–4th leaf-stage were planted to plastic containers (17 × 29 × 9.5 cm) in the growth chamber at 20/15°C (day/night) with a light intensity of 74–81 μmol m^{-2} s^{-1} and a 14-hours photoperiod. The containers were filled with 3 L of 25% standard 'Enshi' solutions. Six plantlets were planted in each container using urethane foam blocks as support. There were four types of nutrient solution viz. renewed, non-renewed, non-renewed with ED at weekly and non-renewed with ED at biweekly intervals. The nutrient solutions were renewed at every week intervals (control). Nutrient solutions of ED treatments were combined in a plastic bucket and then ED was applied. During ED application the nutrient solutions were aerated continuously by air pump (NISSO, Chikara α 4000 SW, Osaka, Japan).

Table 8. Effects of electro-degradation (ED) on the growth of strawberry plantlets in growth chamber bioassay. Culture solutions were treated as renewed, non-renewed, non-renewed with weekly ED and non-renewed with biweekly ED. ED of culture solution were done by electrode used in previous experiment in 2008 (Asao et al. 2008)

Culture solution	Number of leaves*	Crown dia. (mm)	Root length (cm)	Shoot fresh wt. (g)	Shoot dry wt. (mg)	Root dry wt. (mg)
Renewed	8.9 a†	7.2 a	14.1 ab	2.9 a	444.5 a	86.0 ab
Non-renewed	8.0 a	4.8 bc	10.6 b	2.1 a	307.8 ab	66.1 b
ED weekly	3.7 b	4.3 c	7.7 b	0.9 b	171.1 b	45.9 b
ED biweekly	8.2 a	5.9 b	15.8 a	2.4 a	390.6 a	93.9 a

* Parameters measured per plant basis. wt.:weight.
† Values in a column followed by different letter differ significantly by Tukey's test ($P<0.01$; $n = 6$).

The ED was applied for 24 h with electric current adjusted at 2 A. After completion of ED, full amount of Fe3+ was added in the treated nutrient solution and the amount of nutrient solution was adjusted to 3 L by adding standard nutrient solution. During the growth periods the major nutrients (NO^{3-}, $PO4^{3-}$, K^{+}, Ca^{2+}, Mg^{2+}, and Fe^{3+}) content in the culture solution were adjusted as close as possible to initial concentrations. After seven weeks, the number of leaves, crown diameter, root length, shoot fresh weight, shoot dry weight and root dry weight were measured.

The ED of the culture solution of strawberry plantlets grown in growth chamber showed a significant influence on growth parameters (Table 8). The weekly renewed nutrient solution leads a vigorous plant growth while inhibition of vegetative growth of strawberry plantlets resulted due to non-renewed nutrient solution during the entire growth period. This growth inhibition is basically attributed by the accumulation of autotoxic root exudates in the non-renewed culture solution. In previous study, Kitazawa et al. (2005) identified benzoic acid as the most potent growth inhibitor that accumulated in strawberry culture solution. In biweekly ED application, growth was not inhibited rather recovered significantly to renewed nutrient solution except crown diameter. However, ED when applied weekly, the growths were inhibited remarkably as in non-renewed nutrient solution. Shoot weight was found to be the worst affected in weekly ED of non-renewed nutrient solution and it was about 69 and 62% reduction in shoot fresh and dry weight against renewed nutrient solution. Growth retardation in weekly ED culture solution also produced fewer new leaves and shorter root length (about 54 and 45%, respectively) compared to renewed control. The major reason of this drastic growth inhibition in weekly ED culture is low concentration of Fe^{3+} in the degraded solution. In previous study it was found that 24 h ED decreases the Fe-EDT concentration about 10% of the initial (Asao et al. 2008). Therefore in this bioassay the effects of ED treatment were immense in weekly than biweekly intervals with the ED machine used previously. The other possible reasons of this growth retardation and timing of ED were investigated following without plant experiments.

4.3. Electro-Degradation of Nutrient Solution in No Plant Experiment

ED treatment was applied in nutrient solution following without plant experiment. In these experiments 10 L of 25% 'Enshi' nutrient solution were electro-degraded with ED machine used previously for 24 h. The electric current was adjusted at 2 A. During the ED process nutrient solutions were aerated constantly by an air pump. Treated nutrient solutions were sampled at 0, 1, 3, 6, 9, and 24 h. In plastic bottles 25 ml samples were collected for major nutrient analysis following the procedures described earlier. The electric voltage and temperature was recorded from ED machine, electrical conductivity was measured by EC meter (ES-51, Horiba, Ltd., Kyoto, Japan), and pH was also recorded from a pH meter (D-12, Horiba, Ltd., Kyoto, Japan) during each sampling.

Table 9. Electric conditions of 24 h electro-degradation (ED) of nutrient solution by the machine used in previous experiment in 2008 (Asao et al. 2008)

Electrolysis duration (h)	Voltage (V)	Electric Conductivity ($mScm^{-1}$)		pH		Temperature (°C)	
	+*	+	–	+	–	+	–
0	14.9	0.84	0.83	6.95	6.98	14.4	14.3
1	14.3	0.84	0.83	7.20	7.30	15.8	14.5
3	13.9	0.79	0.79	7.36	7.35	18.0	14.7
6	13.6	0.72	0.82	7.26	7.28	20.4	14.5
9	13.3	0.75	0.82	6.86	7.40	22.1	14.5
24	12.7	0.79	0.82	3.13	7.26	22.6	14.2

* (+,–) indicate with or without electrodegradation.

The electrolysis timing and intensity were investigated in without plant experiment (Table 9). ED of nutrient solution was conducted at 2 A for 24 h by the ED machine used previously. After 24 h ED electrical voltage were decreased by 2.2 V whereas the electrical conductivity was not varied with the electrolysis durations. However, the pH of solution was decreased as low as 3.1 after 24 h of ED but it was almost not changed till 9 h ED. This exceedingly low pH after 24 h of ED was another important cause that leads to strong inhibitory effects on strawberry plantlets growth at weekly ED in the growth chamber bioassay (Table 8).

The solution temperature also increase considerably (over 9.1°C form starting) after 24 h ED compared to non ED solution. Changes in concentration of nutrients were determined and it was found that Ca^{2+} decreased (about 66%) considerably after 24 h ED (Figure 5). In addition to decrease in Fe^{3+} concentrations in the ED treated nutrient solution, Ca^{2+} also found to be decreased after 3 h ED. Therefore, ED of nutrient solution for 24 h was associated with some problems, it was not only degradated the allelochemicals but also breakdown the essential nutrients like Ca^{2+} and Fe^{3+}. Considering the very low pH and decrease in Ca^{2+} and Fe^{3+} concentration in the treated nutrient solution a shorter ED duration (2 h) was selected for Wagner's pot hydroponic system for strawberry. Moreover, short electrolysis duration can save electricity and can be feasible in the greenhouse strawberry cultures.

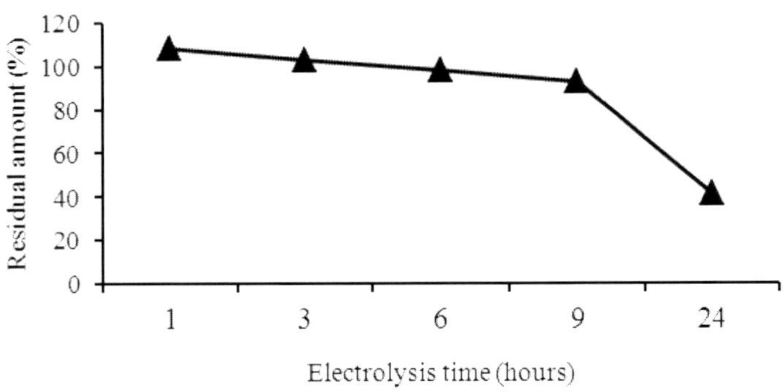

Figure 5. Decrease in Ca^{2+} concentration in the electro-degraded nutrient solution after 24 hours by the machine used in previous experiment in 2008 (Asao et al. 2008).

4.4. Wagner's Pot Hydroponic System of Strawberry Plants

Strawberry plantlets were planted into Wagner's pot on 16 November 2009. At transplanting the plantlets were at 5th – 6th leaf stage, healthy and similar in size. The culture solutions were prepared as renewed at every week, non-renewed entirely, non-renewed with ED at every two weeks and non-renewed with ED at every four weeks intervals. Strawberry plants after reaching active vegetative phase, ED treatments were applied on 11 January 2010. The electric current was adjusted at 2 A and ED was applied for 2 h. The ED machine was modified to investigate its efficacy for degradating autotoxic chemical following a Wagner's pot hydroponic system in the greenhouse (Figure 4).

The modified machine has an electrode having a central core made of ferrite with a surface area of 65.9 cm^2 (anode) which enclosed with cylindrical tube made of titanium with a surface area of 103.7 cm^2 (cathode). The nutrient solution can pass through the electrode where electrodegradation takes place. The machine was coupled with a digital DC power supplier (AD-8735D, AND, Japan). In the previous study, ED of autotoxic chemicals from strawberry root exudates was done by an electrode (cathode) consisting of a cylindrical titanium plate with a 180 cm2 surface area and ferrite stick of 42 cm^2 surface area (anode) (Asao et al. 2008). The previous ED machine requires a separate pump system to flow the nutrient solution during treatment but the modified machine has the facility of attaching pump for flowing treated nutrient solution homogenously through the electrode. Therefore, the modified ED machine was used for degradation of allelochemicals in Wagner's pot hydroponics of strawberry in greenhouse. The nutrient solution was circulated through the pipes for 5 min at 10 min intervals using an automatic pump timer (KS-1500, Iuchi, Osaka, Japan). The nutrient solutions were renewed at every four weeks interval. The concentration of major nutrients were analyzed and adjusted at four weeks interval. The EC, pH and temperature of nutrient solution were measured at each time of treatment application. The date of anthesis were recorded for each plant to check whether there were any influences of ED on flowering of strawberry among the treatments. Pollination was aided by a soft brush at two days intervals. As the plants attained its full vegetative growth phase, leaf length and width were measured on 10 April 2010 to compare growth among the treatments. The fruits were harvested at their maturity. At each harvest fresh weight of fruits were recorded and gathered for final yield comparison among the treatments. Leaf number, leaf length and width, root length, dry weight of leaves, dry weight of crown, and dry weight of root were also recorded at final harvest.

Table 10. Effects of electro-degradation (ED) on the anthesis and vegetative growth of strawberry plants in Wagner's pot hydroponic system. Culture solutions were treated as renewed, non-renewed, non-renewed with ED at two weeks and non-renewed with ED at four weeks intervals

Culture solution	Date of anthesis (month/day)*	Vegetative growth	
		Leaf length (mm)	Leaf width (mm)
Renewed	2/22	222.2	194.4 a†
Non-renewed	2/20	215.4	162.2 b
ED biweekly	2/23	225.7	178.6 ab
ED four weekly	2/26	228.2	198.3 a
	NS	NS	

* Strawberry plantlets were transplanted on 16 November 2009 and anthesis was observed on February 2010.

† Values in a column followed by different letter differ significantly by Tukey's test ($P<0.05$; $n = 10$).

ED method was applied for recovering the autotoxicity in strawberry, which usually occurs in closed hydroponics (Kitazawa et al. 2005). Therefore, a closed hydroponic system was developed in Wagner's pot for treating the strawberry culture solution in the greenhouse. After active vegetative growth strawberry plants were started flowering. There were no significant influences of ED on the anthesis of strawberry plants but comparatively earlier flowering were observed in plants grown in non-renewed nutrient solution (Table 10). Generally plant switches its vegetative growth to reproductive when it experiences some stresses. The synthesis and exudation of allelochemicals enhanced by stress conditions that the plant encounters such as extreme temperature, drought and UV exposure (Pramanik et al.

2000; Inderjit and Weston, 2003). Thus earlier anthesis in strawberry plants grown in non-renewed nutrient solution might be the results of allelochemical stresses which were released from its root. Delayed anthesis was occurred in plants with ED at four weeks intervals. This phenomenon might be due to degradation of allelochemicals (mainly benzoic acid) form culture solution of strawberry. Vegetative growth measurements showed differences among the culture solutions of strawberry (Table 10). Vigorous growth was observed in plants grown in renewed nutrient solution and in non-renewed nutrient solution with ED at four weeks intervals. Although leaf lengths were not differed significantly, the leaf widths were varied among the treatments. Renewed and non-renewed nutrient solution with ED four weekly intervals produces strawberry plants with wider leaves.

Larger leaves provide higher photosynthetic area that produces higher strawberry yield. Whereas non-renewed and non-renewed with biweekly ED nutrient solution produces less width leaves. The results revealed that the leaf sizes were decreased about 21% in non-renewed nutrient solutions than renewed but non-renewed nutrient.

The growth and yield of strawberry plants were significantly influenced by the culture solutions used (Table 11). Renewed and four weekly ED of non-renewed nutrient solution produced higher dry weights of leaf and root than non-renewed and non-renewed with biweekly ED. Leaf number per plants, maximum leaf length and width, root length, crown dry weights were higher in renewed nutrient solution and non-renewed with four weekly ED followed by non-renewed and its biweekly ED. This indicates during autotoxicity, growth in strawberry plants were recovered by ED treatment. In cucumber, it was found that plant growth was inhibited by its own root extract and exudates while improved by removal of these substances from the rhizosphere (Yu and Matsui, 1994; Asao et al. 1998; Yu et al. 2000).

Table 11. Effects of electro-degradation (ED) on the growth and yield of strawberry plants in Wagner's pot hydroponic system. Culture solutions were treated as renewed, non-renewed, non-renewed with ED at two weeks and non-renewed with ED at four weeks intervals

Culture solution	Number of leaves*	Leaf length (cm)	Leaf width (cm)	Root length (cm)	Dry weight of leaves (g)	Dry weight of crown (g)	Dry weight of root (g)	Number of harvested fruits	Fruit yield (g)
Renewed	60.4	32.7	20.0 a†	37.4	47.3 a	2.4	7.5 a	30.8 a	295.0 a
Non-renewed	58.8	28.3	15.9 b	35.2	32.2 b	1.6	5.5 b	26.7 ab	223.5 bc
ED biweekly	58.8	30.5	18.4 a	37.5	35.7 b	2.1	5.7 b	22.5 b	220.6 c
ED four weekly	55.9	30.3	19.7 a	34.4	40.0 ab	2.2	6.0 ab	30.3 a	290.7 ab
	NS	NS		NS		NS			

* Parameters measured on per plant basis.

† Values in a column followed by different letter differ significantly by Tukey's test ($P<0.05$; $n = 10$).

NS:Non significant.

The number of fruits in non-renewed nutrient solution and biweekly electro-degraded solution decreased to 13% and 27%, respectively compared with renewed control. Whereas, there were no significant difference in fruit number between renewed and non-renewed nutrient solution having four weekly ED. The fruit yield of strawberry in non-renewed nutrient solution and its biweekly ED treatment were decreased about 25% compared with control (renewed four weekly). The complete recovery (about 99%) of fruit yield was resulted in plants grown in non-renewed nutrient solution with four weekly ED. In biweekly ED of non-renewed nutrient, Ca^{+2} and Fe^{3+} were broken in addition to degradation of allelochemicals from the strawberry plant roots. Biweekly ED is more intense compared to four weekly and therefore, contain lower concentration of major nutrients like calcium and iron in the culture solution. As a result strawberry plants were found to be retarded in these culture solutions. In previous study, ED of non-renewed nutrient solution, strawberry fruit yield recovered about 71% to renewed control (Asao et al. 2008) and complete yield recovery was not achieved due to decrease in Fe3+ in the nutrient solution. In the present Wagner's pot hydroponics appropriate timing and intensity of ED were applied and therefore complete recovery of fruit yield was obtained.

Table 12. Soluble solid content, citric acid and ascorbic acid content in fruits of strawberry grown in nutrient solution with electro-degradation (ED); fruits harvested at stage I (2010/3/23~2010/4/30), stage II (2010/5/1~2010/5/15), and stage III (2010/5/16~2010/6/8). Culture solutions were treated as renewed, non-renewed, non-renewed with ED at two weeks and non-renewed with ED at four weeks intervals

Culture solution	Soluble solid content (%)			Citric acidity (%)			Ascorbic acid (ppm)		
	Stage I	Stage II	Stage III	Stage I	Stage II	Stage III	Stage I	Stage II	Stage III
Renewed	9.1	9.0	9.8	0.53	0.54	0.61	630.6	583.2	587.2
Non-renewed	9.1	9.4	9.6	0.41	0.46	0.43	632.1	635.0	780.2
ED biweekly	8.6	8.9	9.1	0.35	0.39	0.46	816.9	611.9	602.5
ED four weekly	8.6	8.2	9.1	0.58	0.51	0.62	671.6	662.2	701.4
	NS*	NS	NS	NS	NS	NS	NS	NS	NS

* Non significant.

The quality of strawberry fruits were not differed significantly in the culture solutions either renewed and/or non-renewed (Table 12). Soluble solid content, citric acid acidity, and ascorbic acid content in the strawberry fruits were not varied among the culture solutions used. Therefore, ED of nutrient solution did not left any detrimental effects on the fruit quality in strawberry.

4.5. Mineral Nutrient Content in Strawberry Plants Grown in Nutrient Solution with Electrodegradation

Strawberry plant parts were separated into leaves, crown and roots and kept in a constant temperature oven (DKN 812, Yamato Scientific Co., Ltd. Japan) for 72 h at 80°C. When the dry matter reaches constant weight, it was ground into powder with a mixer machine (National MX-X53, Japan). Samples weighing 0.5 g were mixed with 8 ml of HNO_3 and digested by microwave sample preparation system (ETHOS1, Milestone S.r.l, Bergamo, Italy). After digestion samples were measured up to 50 ml of volumetric flask and then filtered with qualitative filter paper (Advantec Grade no. 131, 185 mm thickness). The filtered sample solutions were analyzed for mineral nutrients by atomic absorption spectrophotometer (Z-2310, Hitachi High Technologies Corporation, Tokyo, Japan).

Table 13. Mineral nutrient content in strawberry plants grown in nutrient solution with electro-degradation (ED). Culture solutions were treated as renewed, non-renewed, non-renewed with ED at two weeks and non-renewed with ED at four weeks intervals

Culture solution	Ca (mg plant^{-1})			Fe (mg plant^{-1})		
	Leaves	Crown	Root	Leaves	Crown	Root
Renewed	0.42 a*	0.029 a	0.063 a	0.062 a	0.0092 a	0.062 b
Non-renewed	0.26 c	0.020 b	0.043 b	0.039 c	0.0052 b	0.042 c
ED biweekly	0.23 c	0.027 a	0.045 b	0.038 c	0.0052 b	0.064 ab
ED four weekly	0.31 b	0.028 a	0.062 a	0.054 b	0.0065 b	0.080 a

* Values in a column followed by different letter differ significantly by Tukey's test ($P<0.01$; $n = 10$).

Calcium and iron contents were determined in the plant parts of strawberry as these two minerals were found to be decreased from the nutrient solution due to ED (Table 13). Calcium contents in non-renewed nutrient solution and its biweekly ED were decreased significantly in all plant parts except in crown compared to renewed and non-renewed nutrient solution with four weekly ED. Calcium uptake in plants grown in non-renewed nutrient solution was inhibited by the accumulated autotoxic chemicals whereas in case of biweekly ED of non-renewed solution, it was due to the breakdown of Ca^{2+} in the nutrient solution by the ED. Iron content in strawberry plants were differed significantly among the culture solution used (Table 13). In leaves and crown Fe^{3+} content were decreased in non-renewed and/or ED nutrient solution while in root it was decreased only in non-renewed nutrient solution. In previous study it was found that 10% Fe-EDTA degradated from nutrient solution due to ED (Asao et al. 2008).

CONCLUSION

Amino acids might have some effects on the growth and yield of strawberry in closed hydroponic system where autotoxicity is a common problem. From the first experiment, it was found that amino acid application left positive effects on the growth and yield of strawberry. Twenty two amino acids were short listed to Ala, Glu, Hyp, Thr, His and Phe on basis of their performance on growth and yield of strawberry in non-recycled hydroponics. In the second experiment, selected amino acids along with GABA were further investigated following Wagner's pot hydroponics. Among the seven amino acids applied, only Glu and Hyp improved their fruit yield over 50%. The *in vitro* culture of strawberry plantlets showed similar results of previous experiments. Therefore, considering the effects of amino acids on growth and fruit yield of strawberry, Glu and Hyp can be used for foliar application in a non-recycled hydroponic culture. Further investigation is necessary to determine the timing and doses of amino acids application for more efficient utilization by the strawberry plants.

The previous study concluded that ED treatment of nutrient solution could mitigate autotoxicity plants in closed hydroponic culture and recovered fruit yield upto 71% of control. However, further studies were suggested to determine the appropriate timing and intensity of ED treatment to maintain the nutrient solution and allow vigorous plant growth. In this present follow up study, we found complete recovery in fruit yield of strawberry plants grown in closed hydroponic system where culture solutions were allowed to ED treatment before reuse. In the growth chamber bioassay strawberry plantlets showed retarded growth in non-renewed and/or its weekly ED whereas it was recovered in biweekly applications. Therefore, electro-degraded nutrient solutions were investigated following without plant experiments. It was found that ED of nutrient solution was associated with degradation of Fe-EDTA (about 10%) and low concentration of Ca2+ (about 66% of the initial) in the treated culture solution. The disadvantages of longer duration ED were decrease in solution pH as low as 3.1 and increase in temperature. Therefore, shorter duration (2 h) ED treatments were applied in the greenhouse experiments following Wagner's pot hydroponics. ED machine used for previous experiments was modified for allowing the treated solution passed continuously through the electrode. Results obtained from Wagner's pot experiment showed that growth and yield (about 99% to renewed control) of strawberry plants were improved in non-renewed nutrient solution having

ED at four weeks interval. Therefore, in practice, application of ED to non-renewed nutrient solution in closed hydroponic system for 2 h at every four weeks intervals would recover autotoxicity in strawberry.

REFERENCES

Abd El-Samad, H.M., Shaddad, M.A.K., Barakat, N., 2011. Improvement of plants salt tolerance by exogenous application of amino acids. *J. Med. Plant. Res.* 5, 5692-5699.

Amin, A.A., Gharib, F.A.E., El-Awadi, M., Rashad, E.S.M., 2011. Physiological response of onion plants to foliar application of putrescine and glutamine. *Sci. Hort.* 129, 353-360.

Asad, S., Arshad, M., Mansoor, S., Zafar, Y., 2009. Effect of various amino acids on shoot regeneration of sugarcane (*Saccharum officinarum* L.). *African J. Biotechol.* 8, 1214-1218.

Asao, T., Hasegawa, K., Sueda, Y., Tomita, K., Taniguchi, K., Hosoki, T., Pramanik, M.H.R., Matsui, Y., 2003. Autotoxicity or root exudates from taro. *Sci. Hort.* 97, 389-396.

Asao, T., Kitazawa, H., Ban, T., Pramanik, M.H.R., 2008. Electrodegradation of root exudates to mitigate autotoxicity in hydroponically grown strawberry (*Fragaria* × *ananassa* Duch.) plants. *HortSci.* 43, 2034-2038.

Asao, T., Kitazawa, H., Ban, T., Pramanik, M.H.R., Matsui, Y., Hosoki. T., 2004a. Search of autotoxic substances in some leafy vegetables. *J. Jpn. Soc. Hort. Sci.* 73, 247-249.

Asao, T., Kitazawa, H., Ushio, K., Sueda, Y., Ban, T., Pramanik, M.H.R., 2007. Autotoxicity in some ornamentals with the means to overcome it. *HortSci.* 42, 1346-1350.

Asao, T., Kitazawa, H., Tomita, K., Suyama, K., Yanamoto, H., Hosokiand, T., Pramanik. M.H.R., 2004b. Mitigation of cucumber autotoxicity in hydroponic culture using microbial strain. *Sci. Hort.* 99, 207-214.

Asao, T., Ohba, Y., Tomita, K., Ohta,. K., Hosoki.T., 1999. Effects of activated charcoal and dissolved oxygen levels in the hydroponic solution on the growth and yield of cucumber plants. *J. Jpn. Soc. Hort. Sci.* 68, 1194-1196.

Asao, T., Tomita, K., Taniguchi, K., Hosoki, T., Pramanik, M.H.R., Matsui. Y., 2000. Effects of activated charcoal supplementation in the nutrient solution on the harvested fruit number of cucumbers grafted on the bloomless rootstock in hydroponics. *J. Soc. High Tech. Agri.* 12, 61-63.

Asao, T., Umeyama, M., Ohta, K., Hosoki, T., Ito, T., Ueda, H., 1998. Decrease of yield of cucumber by non-renewal of the nutrient hydroponic solution and its reversal by supplementation of activated charcoal. *J. Jpn. Soc. Hort. Sci.* 67, 99-105 (in Japanese with English summary).

Barkosky, R.R., Einhellig, F.A., 1993. Effects of salicylic acid on plant-water relationships. *J. Chem. Ecol.* 19, 237–247.

Baziramakenga, R., R.R., Leroux, Simard, R.R., 1995. Effects of benzoic and cinnamic acids on membrane permeability of soybean roots. *J. Chem. Ecol.* 21, 1271–1285.

Baziramakenga, R., Simard, R.R., Leroux, G.D., 1994. Effects of benzoic and cinnamic acids on growth, mineral composition and chlorophyll content of soybean roots. J. Chem. Ecol. 20, 2821–2833.

Blum, U., Shafer, R., Lehmen, M.E., 1999. Evidence for inhibitory allelopathic interactions including phenolic acids in field soils: Concept vs. an experimental model. *Crit. Rev. in Plant Sci.* 18, 673-693.

Bowman, D.C., Paul, J.L., 1992. Foliar absorption of urea, ammonium, and nitrate by perennial ryegrass turf. *J. Amer. Soc. Hort. Sci.* 117, 75-79.

Claparols, I., Santosa, M.A., Torne, M.J., 1993. Influence of some exogenous amino acids on the production of maize embryogenic callus and on endogenous amino acid content. *Plant Cell Tissue Organ Cult.* 34, 1-11.

Comninellis, C.H., Pulgarin, C., 1991. Anodic oxidation of phenol for waste water treatment. *J. Applied Electrochem.* 21, 703–708.

Cruz-Ortega, R., Anaya, A.L., Romos, L., 1988. Effects of allelopathic compounds from corn pollen on respiration and cell division of watermelon. *J. Chem. Ecol.* 14, 71–86.

Das, C., Sengupta, T., Chattopadhyay, S., Setua, M., Das, N.K., Saratchandra, B., 2002. Involvement of kinetin and spermidine in controlling salinity stress in mulberry (Morusalba L. cv. S1). Acta Physiol. Plant. 24, 53-57.

Dashek, W.V., Harwood, H.I., 1974. Proline, hydroxyproline, and lily pollen tube elongation. *Ann. Bot. 38*, 947-959.

Doblinski, P.M.F., Ferrarese, M.L.L., Huber, D.A., Scapim, C.A., Braccini, A.L., and Ferrarese, F.O., 2003: Peroxidase and lipid peroxidation of soybean roots in response to p-coumaric and p-hydroxybenzoic acids. *Brazilian Archives of Biol. Tech.* 46, 193–198.

Feng, Y.J., Li, X.Y. 2003. Electro-catalytic oxidation of phenol on several metal-oxide electrodes in aqueous solution. *Water Res.* 37, 2399–2407.

Fleszar, B., Ploszynka, J., 1985. An attempt to define benzene and phenol electrochemical oxidation mechanism. *Electrochimica Acta* 30, 31–42.

Furuya, S., Umemiya, Y., 2002. The influence of chemicalforms on foliar-applied nitrogen absorption for peach trees. Proceedings of the International Seminar on Foliar Nutrition. *Acta Hort. ISHS* 594, 97-103.

Grewel, D., Gill, R., Gosal, S., 2006. Role of cysteine in enhancing androgenesis and regeneration of indica rice (*Oryza sativa* L.). *Plant Growth Reg*. 49, 43-47.

Hamasaki, R.M., Purgatto, E., Mercier, H., 2005. Glutamine enhances competence for organogenesis in pine apple leaves cultivated *in vitro*. *Braz. J. Plant Physiol*. 17, 383-389.

Holappa, L.D., Blum, U., 1991. Effects of exogenously applied ferulic acid, a potential allelopathic compound, on leaf growth, water utilization, and endogenous abscisic acid levels of tomato, cucumber, and beans. *J. Chem. Ecol*. 17, 865–886.

Hori, H., 1966. Gravel culture of vegetables and ornamentals. Yokendo, Tokyo, Japan. pp. 60-79 (in Japanese).

Inderjit, Weston, L.A. 2003. Root exudates: an overview. In: KROON, H. de and E.J.W. VISSER (eds.): Root ecology. Ecological Studies 168, Springer, Verlag Berlin, Heidelberg, New York, 235–255.

Kieliszewski, M.J., 2001. The latest hype on Hyp-*O*-glycosylation codes. *Phytochem*. 57, 319-323.

Kieliszewski, M.J., Shpak, E., 2001. Synthetic genes for the elucidation of glucocylation codes for arabinogalactan-proteins and other hydroxyproline-rich glycoproteins. *Cell. Mol. Life Sci*. 58, 1386-1398.

Kitazawa, H., Asao, T., Ban, T., Hashimoto, Y., Hosoki, T., 2007. 2,4-D and NAA supplementation mitigates autotoxicity of strawberry in hydroponics. *J. Appl. Hort*. 9, 26-30.

Kitazawa, H., Asao, T., Ban, T., Pramanik, M.H.R., Hosoki, T., 2005. Autotoxicity of root exudates from strawberry in hydroponic culture. *J. Hort. Sci. Biotech*. 80, 677-680.

Koshikawa, K., Yusuda, M. 2003. Studies on the bench culture with closed hydroponic system in strawberry (Part I). *J. Japanese Soc. Hort. Sci*. 72, 394 (In Japanese).

Lea, P.J., Azevedo, R.A., 2006. Nitrogen use efficiency. 1. Uptake of nitrogen from the soil. *Ann. Appl. Bio*. 149, 243-247.

Lea, P.J., Ireland, R.J., 1999. Nitrogen metabolism in higher plants. In: Singh B. K, (Eds.) Plant Amino Acids: Biochemistry and Biotechnology. New York, Marcel Dekker. pp.1-47.

Lyu, S.W., Blum, U., 1990, Effects of ferulic acid, an allelopathic compound, on net P, K, and water uptake by cucumber seedlings in a split-root system. *J. Chem. Ecol*. 16, 2429–2439.

Mazher, A.A.M., Zaghloul, S.M., Mahmoud, S.A., Siam, H.S., 2011. Stimulatory effect of kinetin, ascorbic acid and glutamic acid on growth and chemical constituents of *Codiaeum variegatum* L. plants. American-*Eurasian J. Agric. Environ. Sci*., 10, 318-323.

Mersie, W. Singh, M. 1993. Phenolic acids affect photosynthesis and protein synthesis by isolated leaf cells of velvetleaf. *J. Chem. Ecol*. 19, 1293–1310.

Miller, D.A., 1996. Allelopathy in forage crop systems. Agron. J. 88, 854–859.

Nassar, A.H., El-Tarabily, K.A., Sivasithamparam, K., 2003. Growth promotion of bean (*Phaseolus vulgaris* L.) by a polyamine-producing isolate of Streptomyces griseoluteus. *Plant Growth Regul*. 40, 97-106.

Oka, S. 2002, Development of the labor-saving cultivation techniques by raising the labor-saving cultivars of vegetables (Part I). Bulletin of the National Agriculture Research Centre, Western Region, *Okayama Prefecture*, 13, 26–27 (In Japanese).

Overland, L. 1966. The role of allelopathic substances in the "smother crops" barley. *Am. J. Bot.* 53, 423–432.

Peltonen, J., 1993. Interaction of late season foliar spray of urea and fungicide mixture in wheat production. *J. Agron. Crop Sci.* 170, 296-308.

Penuelas, J., Ribas-Carbo, M., and L. Giles, M., 1996. Effects of allelochemicals on plant respiration and oxygen isotope fractionation by the alternative oxyldase. *J. Chem. Ecol.* 22, 801–805.

Petrova, A.G., 1977. Effect of phytoncides from soybean, gram, chickpea and bean on uptake of phosphorus by maize. In: Grodzinsky, A.M. (eds.): Interaction of Plants and Microorganisms in Phytocenoses. Kiev: Naukova Dumka, 91–97 (in Russian with English summary).

Pramanik, M.H.R., Nagai, M., Asao, T., Matsui, Y., 2000. Effects of temperature and photoperiod on phytotoxic root exudates of cucumber (*Cucumis sativa*) in hydroponic culture. *J. Chem. Ecol.* 26, 1953–1967.

Putnam, A.R. 1985b. Weed allelopathy. In: Duke, S.O. (ed.): Weed physiology: Reproduction and Ecophysiology. CRC Press, Boca Raton FL, 131–155.

Putnam, A.R., 1985a. Allelopathic research in agriculture: Past highlights and potential. In: Thompson, A.C. (ed.): The Chemistry of allelopathy: Biochemical interactions among plants. *Am. Chem. Soc.*, Washington, D.C. USA, 1–8.

Rajagopal, V., Sinha, S.K., 1980. Influence of exogenously supplied proline on relative water content in wheat and barley. *Indian J. Exp. Biol.* 18, 1523-1524.

Rao, A.M., Sree, K.P., Kishor, P.B.K., 1995. Enhanced plant regeneration in grain and sweet sorghum by asparagine, proline and cefotaxime. *Plant Cell Rep.* 15, 72-75.

Rice, E.L. 1984, Allelopathy. 2nd Ed. Academic Press, Orlando, Florida, 422.

Rohn, S., Rawel, H.M., Kroll, J., 2002. Inhibitory effects of plant phenols on the activity of selected enzymes. *J. Agri. Food Chem.* 50, 3566–3571.

Singh, H.P., Batish, D.R., Kohli, R.K., 1999. Autotoxicity: concept, organisms and ecological significance. *Crit. Rev. Plant Sci.* 18, 757–772.

Skokut, T.A., Manchester, J., Schaefer, J., 1985. Regeneration in alalfa tissue culture. *Plant Physiol. 79*, 579-583.

Takeuchi, M., Arakawa, C., Kuwahara, Y., Gemma, H., 2008. Effects of L-pro foliar application on the quality of 'Kosui' Japanese pear. *Acta Hort.* 800, 549-554.

Takeuchi, T., 2000. The nourishment uptake of strawberry cultivar 'Akihime' in rockwool hydroponics with a nutrient solution circulating system. *Bulletin of Shizuoka Agricultural Experimental Station* 45, 13–23 (In Japanese with English summary).

Tang, C.S. and Young, C.C. 1982. Collection and identification of allelopathic compounds from the undisturbed root system of bitalta limpograss (*Helmarthria altissima*). *Plant Physiol.* 69, 155–160.

Thakur, P.S., Rai, V.K., 1985. Exogenous supplied of amino acids and water deficits in *Zea mays* cultivar. *Biol. Plant.* 27, 458-461.

Vasilas, B.L., Leeg, J.O., Wolf, D.C., 1980. Foliar fertilization of soybeans: Absorption and translocation of ^{15}N-labeled urea. *Agron. J.* 72, 271-275.

Waller, G.R., Nowacki, E.K., 1978. Alkaliod, Biology and Metabolism in Plants. Plenum Press, New York. pp. 5-247.

Weir, T., Park, S.W., Vivianco, J.M., 2004. Biochemical and physiological mechanisms mediated by allelochemicals. *Curr. Opin. Plant Biol.* 7, 472-479.

Yaronskaya, E., Vershilovskaya, I., Poers, Y., Alawady, A.E., Averina, N., Grimm, B., 2006. Cytokinin effects on tetrapyrrole biosynthesis and photosynthetic activity in barley seedlings. *Planta. 224*, 700-709.

Yu, J.Q., Matsui, Y. 1994. Phytotoxic substances in the root exudates of *Cucumis sativus* L. *J. Chem. Ecol*. 20,21–31.

Yu, J.Q., Shou, S.Y., Qian, Y.R., Hu, W.H., 2000. Autotoxic potential in cucurbit crops. *Plant Soil* 223, 147–151.

In: New Developments in Allelopathy Research
Editor: Julia E. Price
ISBN: 978-1-63483-390-5

Chapter 4

ALLELOPATHY OF GLOBALLY SUCCESSFUL INVADER *CONYZA CANADENSIS* L.

L. Djurdjević[*], P. Pavlović, G. Gajić and M. Mitrović
Institute for biological research Siniša Stanković, University of Belgrade, Serbia

ABSTRACT

The globally successful invader *C. canadensis* is the first species to colonise the sandy levees beside the rivers into new areas and, owing to high levels of seed production and in particular its allelopathic effect, it completely dominates the other species in the ruderal phytocoenosis. The content of total phenolics and phenolic acids varies, following the order: vegetative plant parts > dead plant parts > sandy soil under *C. canadensis*. Water leachate and soils inhibited seed germination and seedling growth of the neighbouring herbaceous plantsto varying degrees: vegetative parts > dead parts > sandy soil, which is directly related to the content of total phenolics and phenolic acids in them. Secondary metabolites, above all phenolic compounds released from its vegetative organs, reach the sand under it, where they have a negative effect on seed germination and seedling growth of neighbouring herbaceous plants, reducing their abundance and cover.

The dynamics of allelopathically significant phenolic compounds in *C. canadensis* L. tissues and in the sandy soil have a marked seasonality and coincide with certain phases in plant development. The highest amount of phenolics in the tissue was measured during the phase of plant elongation and intensive growth, and again during the phase of fruit abscission and plant decline. It can be seen that the level of free acids in the soil under *C. canadensis*, increases dramatically between March, when the plants are in the form of rosettes, 1-2 cm in diameter, and July, the beginning of flowering. In these months free phenolic acids account for as much as 73.7% of the total free phenolics in the soil. The percentage of soil bound phenolic acids gradually increases between April, when the plants are in the form of larger rosettes, 5-10 cm in diameter, and August (in full flower and beginning of fruiting), when their maximum level (43.5%) is reached. Phenolic acids can serve as reliable bioindicators of phenolics metabolism and turnover, starting from their biosynthesis in plant tissues to their incorporation into stable fractions of soil

[*] Bulevar Despota Stefana 142, 11060, Belgrade, Serbia. E. mail: kalac@ibiss.bg.ac.rs.

humus. Also, they show the level of improvement in substrate quality for further plant colonisation in comparison to sands. An important role in the spread of this invasive species into new areas and its domination over native species is played by phenolic compounds as the most abundant group of allelochemicals. These exert regulatory roles in many processes, not only in plants themselves, but as well in the ecosystem as a whole.

Keywords: allelopathy, *Conyza canadensis*, biodiversity, phenolics, sand, degraded habitats

INTRODUCTION

Allelopathy has been explored as a potential driver of plant invasion in the context of the novel weapon hypothesis (NWH), part of which emphasizes that some invasive plants may have stronger allelopathic impacts on other species in introduced areas than in native areas because the former may not be as adapted to the specific allelochemcials of invaders as plants in the native range (Callaway and Ridenour, 2004). There is increasing evidence that many plant invaders interfere with native plants through allelopathy. This allelopathic interference may be a key mechanism of plant invasiveness (Bousquet-Mélou et al., 2005, Dorning and Cipollini, 2006, Callaway and Maron, 2006, Inderjit et al., 2008, Thorpe et al., 2009, Pisula and Meiners, 2010, Murrell et al., 2010, McEwan et al., 2010; Kim and Lee, 2011). Many results suggest that the inherently stronger competitive and allelopathic effects of invasive plants on plants in non-native areas than on plants from its native range may contribute to its invasive success. Invasive plants excel in their new ranges because they produce novel metabolites to which native species possess little resistance (Ni et al., 2010, Thorpe et al., 2009, Barto et al., 2010, Orr et al., 2005 and Hierro and Callaway, 2003).

Rice divided all the inhibitory compounds that are important in allelopathy (phytoncides and kolines), which are secondary metabolites in higher plants, into 15 groups of compounds (Rice, 1974). The group of allelochemicals that has been studied most are phenolic compounds, which are universally distributed in all plants, in all plant organs, as free and bound forms - of which several thousand exist. They are part of the composition of lignin, the most abundant polymer in plants or are bound to cell wall polysaccharides (Harborne, 1980, Bate-Smith, 1969, Saunders and McClure, 1976, Kögel, 1986, Whitmore, 1976, Djurdjević et al., 1998, Djurdjević et al., 2004, Djurdjević et al., 2005a, Djurdjević et al., 2005b and Djurdjević et al., 2011). Phenolic compounds enter soil through being released from aboveground plant parts by precipitation, through glandular trichomes, through active leaching via roots, and through the microbial degradation of dead plant remains, mostly due to the degradation of lignin (Alexander, 1977; Tang and Young, 1982; Whitehead et al., 1983; Sterling et al., 1987; Grodzinsky et al., 1987; Carballeira and Reigosa, 1999; Reigosa et al., 1999a; Kögel-Knabner, 2002; Hyder et al., 2002a, b; Djurdjević, 1991; Djurdjević and Oberan, 1998; Djurdjević et al., 1997, 1999, 2000, 2003).

In terms of their abundance and primary productivity, the dominant species in plant communities are of great importance as the main source of phenolics, and therefore have a significant effect on plant–litter–soil interactions. Soil fertility is influenced by phenolics from dominant plant or microbial sources not only because they are important precursors of stable soil humic substances, but also for their effects on soil nutrient dynamics (N, P, K, Mn, Fe, Cu, etc.), pH, ion-uptake, the rate of litter decomposition, the composition and activity of soil microbial communities, soil quality, soil C and N cycling, macro- and micronutrient availability, soil aggregation, plant-water relations and soil respiration (Djurdjević et al., 2009). Phenolic toxins originating from dominant plants and accumulating in soil can lead to the inhibition of seed germination, as well as of root and seedling growth, photosynthesis, water and ion uptake of target plants, glyoxylic enzyme activities, etc. As a consequence of the negative effects of phenolics originating from the dominant species, there is a reduction in the abundance of the other species in a phytocoenosis or their complete disappearance, hence disturbing the existing biodiversity (Muller, 1966; Lodhi and Rice, 1971; Rice, 1974; Rice, 1979; Lodhi, 1975; Mersie and Singh, 1988; Lyu and Blum, 1990; Li et al., 1992a, b; Aliotta et al., 1993; Barkosky and Einhellig, 1993; Souto et al., 1995; Reigosa et al., 1999b; Barkosky et al., 2000; Muscolo and Sidari, 2006; Hussain and Reigosa, 2011; Hussain et al., 2011; Timsina et al., 2011; Djurdjević et al., 2004, 2005a, 2008,2011).

One of the globally successful invaders is *Conyza canadensis* L. (horseweed), originating from North America. Now it is widespread as a neophyte across almost the whole world (a cosmopolite). It was introduced into Europe in the 17th century and was first noted in Serbia in 1874 and in the vicinity of Belgrade in 1878 around houses, in gardens, along roadsides, in sand and fields (Pančić, 1874 and Pančić, 1878). In Serbia, it is a naturalized species, which spreads from ruderal to arable land (Gajić, 1975; Stjepanović-Veseličić, L., 1979; Jovanović, 1994; Vrbničanin et al., 2004; Stojšić and Dinić, 2009; Stojšić, et al., 2007; Vrbničanin et al., 2008; Jarić, 2009) and plays a significant role in pioneer communities in degraded habitats (Djurdjević et al., 2011; Zaplata et al., 2011).

C. canadensis occupies large areas of the sandy levees next to the Danube and Tamiš rivers (Serbia), dominating the other species in the phytocoenosis due to its aboveground mass and abundance. Due to the major role it plays in the formation of pioneer communities, we undertook allelopathic research into the domination of this species (Figure 1). The aim of this study was the quantification of total phenolics and phenolic acids as the main allelochemicals in the vegetative and dead plant parts of *C. canadensis* and in the surface sandy soil layer, respectively. The allelopathic potential of this species was studied through observing seed germination and seedling growth of two grasses that grow beneath it. The role of this invasive pioneer species and its secondary metabolites in soil formation on the barren sandy levees is discussed.

This is the first work to study the seasonal dynamics of phenolics in the tissue of the invasive dominant species *C. canadensis* and the soil under it. Samples of plant material (aboveground plant parts) and soil were taken once a month in order to monitor the seasonal dynamics of phenolics, as follows: March (young plants in the form of rosettes, 1-2 cm in diameter), April (young plants in the form of rosettes, 5-10 cm in diameter), May (beginning of elongation, young plants, 10-15 cm long), June (phase of intensive plant growth, plants up to 1.5 m in height), July (beginning of flowering), August (in full flower and beginning of fruiting, plants up to 2 m in height), September (fruit abscission, the beginning of plant decline) (Figure 2). It is believed that through taking samples at monthly intervals throughout the development of *C. canadensis* from the rosette phase to that of seed abscission and plant decline, we will have a complete insight into the dynamics of phenolic compounds formed in the plant tissue and their translocation into the soil under it, as opposed to taking a sample once a year or on just few occasions during the vegetative season. This is of crucial importance for the study of the allelopathic capacity of this invasive species, and its effect on the other species in the phytocoenosis, and its survival in newly-invaded areas away from its native one.

PHYTOCOENOLOGICAL INVESTIGATIONS

C. canadensis L., a cosmopolitan agricultural and ruderal weed, occupies large areas of the sandy levees at the confluence of the Tamiš and Danube rivers as a pioneer species, forming dense microcomplexes and dominating the other herbaceous species in the ruderal phytocoenosis with meadow elements due to its aboveground mass and abundance in Serbia. This species is among the first to colonise the bare sandy terrain and plays an important role in vegetation succession on the sands and increasing the biodiversity. At the beginning of August when samples were collected, horseweed was in the final stage of flowering, fruiting and vigorous growth. Aside from its allelopathic influence, the expansiveness and vitality of this species can be seen through the high density of the population, owing to high levels of seed production and the height of the plants (120-180 cm). In addition to the dominant *C. canadensis*, whose abundance and cover was 75%, a further 25 herbaceous plant species were also found, but their abundance and cover were significantly lower (Table 1, Figures 1, 2).

Figure 1. The ruderal phytocoenosis with *C. canadensis* domination.

C. canadensis quickly colonize the barren sandy levees beside rivers or abandoned areas without a vegetation due to its biological and ecological characteristics, i.e., high levels of seed production, and fast growth of the stems. For example, Regehr and Bazzaz (1979) found that the number of flowers per plant, and therefore the total seed production, is proportional to stem height of *C. canadensis*. A plant 40 cm tall produces approximately 2000 seeds, while a 1.5 m tall plant produces approximately 230,000 seeds. Bhowmik and Bekech (1993) reported that plants of *C. canadensis* grown at a density of 10 plants m^{-2} in a no-till, cropless field produced approximately 200,000 seeds per plant. According to the results of Bruce and Kells (1990), estimated soybean yield losses were 83% from 150 plants m^{-2} of *C. canadensis* in a no-till cropping system, averaged over a 2 year period, which ties in with our results on the low abundance and cover of herbaceous plants under this species.

Table 1. Phytocoenotic releve data of plant community with *C. canadensis* domination

Plant species	Abundance
Conyza canadensis L.	8
Glechoma hederacea L.	1
Thlaspi perfoliatum L.	*1*
Anchusa officinalis L.	1
Stellaria neglecta Weihe	1
Sonchus oleraceus L.	1
Crepis biennis L.	1
Stenactis annua (L.) Nees.	1
Dactylis glomerata L.	1
Trifolium repens L.	1
Roripa silvestris L.	1
Stellaria media (L.) Vill.	1
Lamium purpureum L.	1
Rubus caesius L.	1
Galium aparine L.	1
Taraxacum officinale F. H. Wigg.	1
Capsellabursa-pastoris (L.) Medik.	1
Veronica hederifolia L.	1
Bromus sterilis L.	1
Poa pratensis L.	1
Artemisia vulgaris L.	1
Oenothera biennis L.	1
Melilotus albus Medik.	1
Cirsium arvense (L.) Scop.	1
Galega officinalis L.	1
Lactuca seriola L.	1

For each plant species, the numbers represent an estimation of its abundance and cover in the examined community, expressed in numbers on a scale of 1-9, equivalent to abundance of 1-100%: 1 (1%), 2 (2%), 3 (5%), 5 (25%), 7 (50%), 8 (75%), and 9 (100%) according to Westhoff and van der Marrel (1973).

Similar to the barren sandy levees beside rivers, which *C. canadensis* is the first to colonise, in the abandoned fields of Indiana, Root and Wilson (1973) found that the fields were dominated by the winter annuals *Conzya annuus* and *C. canadensis* during the first year after abandonment. However, after 9 years, these species disappear or their significance decreases. Similar data on the role of *C. canadensis* as a pioneer weed species and its domination during the first few years is given by Kapustka and Moleski (1976). As a weed species, *C. canadensis* is found in 'cultivated fields,' along roadsides and railways, and in abandoned fields (Weaver, 2001).

Figure 2. Some phases in the development of *C. canadensis*: A. Young plants in the form of rosettes, 5-10 cm in diameter; B. Beginning of elongation, young plants, 15-20 cm long; C. Full flowering and beginning of fruiting, plants up to 2 m in height; D. The flowers and fruits.

(http://cals.arizona.edu/yavapaiplants/imageviewer.php?genus=Conyza&species=canadensis&imagenum=6; http://www.plantarium.ru/page/image/id/127196.html; http://www.afleurdepau.com/Flore/asteraceae/conyza-canadensis/x.htm)

Phenolic Compounds in the Vegetative and Dead Parts of *C. canadensis*

The results of measuring the content of total phenolics (free and bound forms) in the vegetative and dead plant parts of *C. canadensis* are given in Table 2. Large amounts of both free and bound forms of total phenolics accumulated in the tissues of the vegetative plant parts. When at the stage of vigorous growth, the vegetative parts of the plant contain levels of free and bound forms of total phenolics that are several times higher than those in the previous year's dead stems. The content of total free phenols in dead plant tissue is only 6% of that measured in vegetative tissue, but with bound forms the level is 65%.

In the tissue of the vegetative parts of *C. canadensis*, we detected 5 phenolic acids: two are derivatives of cinnamic acid (*p*-coumaric and ferulic), and three of benzoic acid (*p*-hydroxybenzoic, vanillic and syringic). The derivatives of cinnamic acid (*p*-coumaric and ferulic) predominate in terms of amount. In comparison to the dead plant tissue, the vegetative parts of the plant contain higher amount of free *p*-coumaric, ferulic and vanillic acid (Table 3). In terms of the ratio of free and bound forms of phenolic acids, the levels of bound forms are higher in both the vegetative and dead tissue. In the vegetative tissue, free acids make up only 2.4% of the total free phenols (31,002 µg/g) due to the presence of other phenolic compounds in a free state, but in the dead tissue they make up as much as 9.74% because the remaining phenolic compounds, those that are not phenolic acids, are washed away by precipitation and get into the soil.

In the tissue of the vegetative and dead plant parts of *C. canadensis*, we detected 5 phenolic acids (*p*-coumaric, ferulic, *p*-hydroxibenzoic, vanillic and syringic). Similarly to our results, gallic acid, vanillic acid, syringic acid and catechol were isolated from the stems of *C. canadensis* (Shaukat et al., 2003). According to our earlier allelopathic research, the aboveground parts of various dominant herbaceous plants contain 1,333-51,218 µg/g of total free phenolics and 1,000-19,718 µg/g of bound phenolics respectively (Djurdjević et al., 1998, 2004, 2005b). The content of phenolic acids is species-dependent, ranging from 3.20-6.79 mg/g. *p*-Coumaric, ferulic, vanillic, caffeic, syringic and *p*-hydroxybenzoic acid were the most frequently found phenolic acids in the stems and roots of many herbaceous plants (Rice and Pancholy, 1973; Rice, 1974; Whitehead et al., 1983; Theodoru et al., 1987; Lowry et al., 1993; Sené et al., 2001; Djurdjević et al., 1998, 2004, 2005b). When *C. canadensis* decline, the high levels of free phenols that have accumulated in the tissue (31,002 µg/g) are quickly washed away by precipitation and get into the soil, inhibiting seed germination and seedling growth of other plants. In the vegetative parts of *C. canadensis*, there are several times fewer bound forms than free and they become available to the other plants in the phytocoenosis through microbial lignin degradation following the death of tissue (Alexander, 1977; Lewis and Yamamoto, 1990).

Table 2. Content of total phenolics in *C. canadensis* plant parts and sandy soil

		Horseweed parts		Sandy soil	
(µg/g)		Dead	Vegetative	Control soil	Soil under horseweed
	free	1,877.58	31,002.00***	4.702	79.60***
Total		± 158.81	± 1,000	± 0.41	± 7.60
phenolics	bound	3,732.00	5,691.00***	26.65	191.79***
		± 373.15	± 564.00	± 3.09	± 18.96
	Total	5,609.58	36,693.00	31.35	271.39

Compared are: (a) dead/vegetative horseweed parts; (b) control soil/soil under horseweed; (ANOVA, ns-not significant; *** $p < 0.001$; n = 5). Means ± S. D.

PHENOLICS IN SOIL UNDER *C. CANADENSIS*

The soil at the levee can be categorised as young sandy soil in the first phases of pedogenesis, in which the dominant species *C. canadensis* has an important role to play, thanks to the decomposition of its dead organic remains and/or the compounds released during vegetation by its roots and aboveground parts. In this soil, 79.60µg/g of total free phenolics and 191.79 µg/g of bound phenolics were measured. As with completely formed soil types, here too there is twice the level of phenolic acids bound in the soil structure than free forms. The content of total phenolics in the control soil without plants is several times lower (Table 2). The levels of the 5 phenolic acids detected in the soil where *C. canadensis* dominates are in the range 0.68-12.52 µg/g for free forms and 10.71-30.44 µg/g for bound forms. Free phenolic acids make up 60% of the total free phenolics in the soil, whereas bound acids amount to 49% of the total bound phenolics (Table 3). The content of phenolic acids in the control soil is several times lower. The amount of the derivatives of cinnamic acid (*p*-coumaric and ferulic) and benzoic acid (*p*-hydroxibenzoic, vanillic and syringic) are almost the same (65.51 µg/g and 76.74 µg/g. Although the dominant *C. canadensis* creates high levels of free phenols (30 mg/g), only an insignificant amount (0.6%) are bound to sand particles because of their low adsorption capacity when compared to soil particles and so they are washed away by precipitation into deeper layers and hence become inaccessible to plants.

Table 3. Content of phenolic acids in *C. canadensis* plant parts and sandy soil

		Horseweed parts		Sandy soil	
(µg/g)		Dead	Vegetative	Control soil	Soil under horseweed
	free	31.87	381.83***	0.13	11.84***
p-Coumaric		± 0.87	± 38.25	± 0.02	± 1.10
	bound	279.56	293.59^{ns}	2.25	10.71***
		± 22.97	± 28.60	± 0.35	± 1.24
	free	12.47	136.57***	0.28	12.52***
Ferulic		± 1.14	± 12.86	± 0.002	± 1.05
	bound	311.49	951.62***	1.40	30.44***
		± 31.00	± 95.00	± 0.17	± 2.80
	free	53.38***	10.13	-	0.68***
p-Hy. benz.		± 5.32	± 1.10	-	± 0.06
	bound	60.73	80.08^{ns}	0.23	19.61***
		± 6.08	± 8.05	± 0.03	± 1.74
	free	55.57	179.49***	0.19	11.58***
Vanillic		± 3.95	± 16.97	± 0.02	± 0.67
	bound	134.30	564.66***	1.21	16.02***
		± 13.85	± 48.35	± 0.45	± 1.27
	free	29.73**	22.41	0.13	11.42***
Syringic		± 3.09	± 1.78	± 0.01	± 0.93
	bound	143.38***	88.62	0.25	17.43***
		± 14.38	± 8.11	± 0.04	± 1.45

Compared are: (a) dead/vegetative horseweed parts; (b) control soil/soil under horseweed; (ANOVA, ns-not significant; ** $p < 0.01$; *** $p < 0.001$; n = 5). Means ± S. D.

In the soil of the floodplain forest *Fraxino angustifoliae-Quercetum roboris* with the autumnal truffle (*Tuber macrosporum* Vitt.) in Serbia, which is situated 20 meters from the levee, the content of total free phenolics was almost the same (74.06 µg/g) as that measured in the soil under *C. canadensis*. However, the content of bound phenolics is 8 times higher, which points to the significantly longer pedogenetic process in the oak-ash community in relation to the pioneer community dominated by *C. canadensis* (Djurdjević et al., 1999). The content of total phenolics in the sandy soil under *C. canadensis* was several times lower in comparison to the other natural soils. Thus, the content of total phenolics in different soils is as follows: in forest-peat bog 6561.3 µg/g (Djurdjević et al., 2003), in mixed Austrian and Scots pine forest 2678.5 µg/g (Djurdjević et al., 1998), in beech-fir forest 1770 µg/g (Djurdjević et al., 2004), in inundate pedunculate oak forest 1589.6 µg/g (Djurdjević et al.,

1999), in virgilius oak forest 708.9 μg/g (Djurdjević et al., 1998), in forest-steppe sandy soil 598.8 μg/g (Djurdjević et al., 2000), and 185.16 μg/g of total phenolics in the zone with the highest vegetation cover (80%), on fly ash deposit after 7 years of revegetation (Djurdjević et al., 2006). In the sandy soil control samples, the content of total phenolics was several times lower. Their presence in sandy soil under *E. canadensis*, although less than in the natural soils, indicated the initiation of pedogenetic processes, processes of organic matter transformation, and humus formation. On the other hand, the presence of phenolics as secondary plant metabolites points to the influence of dominant plants in the overbuilding of substrate such as sand deposits. Organic plant matter existing in the soil is of great importance for the formation of soil humus, which represents the product of biological soil activity and contains mixtures of different organic compounds. In addition to humic acids, humus also contains different peptides, carbohydrates, metals and phenolic compounds (Felbeck, 1971; Haworth, 1971). Phenolic compounds make up an important component of the humus layer of different soil types (Guenzi and McCalla, 1966; Hennequin and Juste, 1967; Katase, 1981a, b; Kögel, 1986; Kuiters and Denneman, 1987, Djurdjević et al., 2000, 2003, 2004, 2005a, b, 2006, 2009; Banach-Szott, 2007).

EFFECTS OF VEGETATIVE AND DEAD PLANT PARTS OF *C. CANADENSIS* AND SANDY SOIL

The results of studying the influence of water leachate from the dead and vegetative parts of *C. canadensis* on seed germination and seedling growth of *Dactylis glomerata* and *Trifolium repens*, which grow under it, but with only a small number of individuals, are given in Table 4. The number of germinated seeds of both species is significantly lower in the leachate from the vegetative plant parts, which is directly connected to the high content of total free phenolics in them. Leachate from the vegetative parts of *C. canadensis* is more inhibitory to the germination of *T. repens* seeds than those of *D. glomerata*. The leachates of the vegetative and dead plant parts of *C. canadensis* had an inhibitory effect on the growth of both the aboveground parts and roots of both the test plants. Vegetative plant parts acted as a strong inhibitor of roots and the elongation of aboveground parts in comparison to dead plant parts. The surface layer of soil (0-10 cm) under *C. canadensis* is equally inhibitory for seed germination of both test plants. Horseweed soil had a more inhibitory effect on the elongation of the aerial parts and roots of *T. repens* in comparison to *D. glomerata* (Table 5).

We found inhibitory effects of water extracts of vegetative and decayed plant parts of *C. canadensis* on seed germination and seedlings growth of test plants. This is in accordance with earlier findings in which the aqueous extract of the vegetative and decaying shoots of *C. canadensis* in sandy-loam inhibited the germination and root and shoot growth of all six test species (Shaukat et al., 2003; Xingxiang et al., 2009). The phytotoxicity of *C. canadensis* varied in the three soil types that differed in soil texture. The greatest activity was recorded in sandy soil. This is presumably due to the fact that this soil has a very small proportion of coloidal material (least clay content) and hence the adsorption of the allelochemicals is minimal (Shukat et al., 2003). Our data confirms these findings on the high phytotoxicity of the dominant species *C. canadensis* on the sandy levee. In additionthe dominant horseweed creates high levels of free phenols but only an insignificant amount are bound to sand particles because of their low adsorption capacity when compared to soil particles, and so they are washed away by precipitation into deeper layers and hence become inaccessible to plants. Because of this, the amount of bound phenols in the soil at the levee is small. This soil can be categorised as young sandy soil in the first phases of pedogenesis when compared to soils that are older in terms of pedogenesis. Invasive species can alter the biogeochemistry of ecosystems, that secondary metabolites released by invasive species may play important roles in soil chemistry as well as plant-plant and plant-microbe interactions (Weidenhamer and Callaway, 2010).

Table 4. Effects of water extracts from dead and vegetative plant parts of *C. canadensis* on seed germination and seedling growth of *Dactylis glomerata* and *Trifolium repens*

Plant species	Treatment	Seed germination %	Length of aerial part (%)	Length of root (%)
D. glomerata	Control	83.15(± 4.00)[a]	100	100
	Dead	53.11(± 4.20)[b]	37.67(± 2.26)[b]	49.63(± 4.61)[b]
	Vegetative	6.14(± 0.49)[b]	27.05(± 3.10)[b]	22.47(± 2.62)[b]
T. repens	Control	79.81(± 4.66)[a]	100	100
	Dead	41.90(± 4.56)[b]	59.39(± 6.11)[b]	38.22(± 4.02)[b]
	Vegetative	4.63(± 0.57)[b]	36.67(± 2.62)[b]	31.94(± 2.66)[b]

[a] Numbers in parentheses are standard deviations. Control = distilled water.
[b] Statistically significant difference, $p < 0.001$ (t-test), n = 500.

Table 5. Effects of sandy soil (0-10 cm) under *C. canadensis* on seed germination and seedling growth of *Dactylis glomerata* and *Trifolium repens* test plants

Plant species	Treatment	Seed germination	Length of aerial part (%)	Length of root (%)
D. glomerata	Control soil	84.36(± 4.40)[a]	100	100
	Horseweed soil	60.01(± 3.40)[b]	73.14(± 2.96)[b]	69.55(± 4.81)[b]
T. repens	Control soil	80.12(± 4.78)[b]	100	100
	Horseweed soil	58.27(± 3.14)[b]	58.30(± 3.68)[b]	58.95(± 3.97)[b]

[a] Numbers in parentheses are standard deviations. Control soil = sandy soil without plants.
[b] Statistically significant difference, $p < 0.001$ (t-test), n = 500.

When studying vegetation succession in abandoned fields in Oklahoma and Kansas (USA), Rice mentions *C. canadensis* as a pioneer weed species along with several other vigorous species whose domination lasts for 2-3 years. He established that the dominant species have an inhibitory effect on other species, among which is *C. canadensis*. The dead leaves of *Helianthus annuus* inhibit *C. canadensis* seed germination (71-87%) and seedling growth (50%). They also have an inhibitory effect on individuals of their own species, forming phenolic compounds, which later accumulate in the soil in toxic concentrations (Rice, 1974, 1979). He did not study the domination of *C. canadensis* over the other species during vegetation succession. Our results also support the research of Chou and Lee (1991), who established that the aqueous extracts of *Miscanthus transmorrisonensis* Hayata displayed significant phytotoxic effects on the seed germination and radicle growth of the plants tested. In addition, rhizosphere soils under *M. transmorrisonensis* also exhibited significant phytotoxicity, indicating that allelopathic interaction was involved. Some responsible phytotoxic phenolics, namely *p*-coumaric, ferulic, vanillic, protocatechuic, o hydroxyphenylacetic, and m-hydroxyphenylacetic acids, and 4-hydroxycoumarin and phloridzin, were identified.

Table 6. The inhibitory influence of phenolic compounds on some physiological process in plants

Phenolics	Physiological process	Inhibitory concentrations	Source
p-Hydroxybenzoic, salycilic, and caffeic acid	Plant-water relationships, chlorophyll fluorescence, stomatal conductance, growth	0.25 mM, 0.5 mM, and 1.0 mM	Barkosky, et al. (2000)
Sinapinic, syringic, vanillic, ferulic, *p*-coumaric, chlorogenic, gallic, gentisic, protocatechuic, *p*-hydroxybenzoic, *trans*-cinnamic acids, eucalyptol, quercetin, vanillin, syringaldehyde, rutin, 2-benzoxazolinone, protocatechualdehyde, tyrosol, juglone, and L-mimosine	Seed germination, root growth, hypocotyls length, root morphology of alfalfa, barnyardgras and weed species, leaf expansion, seedling growth	20 -700μM, 10^{-2}M, 10^{-3}M, 10^{-4}M 10^{-5}M, 10^{-6}M 0.05-0.25 μM/g soil	Chon et al. (2002), Reigosa and Pazos-Malvido(2007), Blum (1996).
Scopoletin	Growth, CO_2 exchange rates	10^{-4}M, 10^{-3}M	Einhellig et al. (1970).
Vanillic and *p*-hydroxybenzoic acids	Radish and grain sorghum germination and sorghum root and shoot elongation	2.5 x10^{-3}M, 5x10^{-3}M	Einhelling and Rasmussen (1978)
Benzoxazolin-2(3*H*)-one, *p*-hydroxybenzoic acid and cinnamic acid	Growth, leaf water relations, PSII photochemistry, non-photochemical fluorescence quenching, and heat energy dissipation	1.5 mM	Hussain and Reigosa (2011)
Ferulic acid	Net P, K, and water uptake	0.5 mM	Lyu and Blum (1990)
p-Coumaric, ferulic, *p*-hydroxybenzoic, vanillic, salicylic and chlorogenic acids	Seed germination and growth of grain sorghum and birch	10-^{3}M and 5 × 10-^{4}M	Rasmussen and Einhelling (1978)
o-Hydroxyphenyl acetic, ferulic and *p*-coumaric acids	Chlorophyll, porphyrin and chlorophyllide contents, and chlorophyllase a and b activities.	20, 50, 100, 200 ppm	Yang, et al. (2002, 2004)
Coumarin and phenylpropanoids	Inhibition of radish seed germination and root growth	2×10-4M	Aliotta et al. (1993)
Ferulic, *p*-coumaric, vanillic, cinnamic *p*-Hydroxybenzoic acid and coumarin	Seed germination, root and shoot length, fresh and dry weight, chlorophyll, carotenoids and anthocyanin content, guaicol peroxidase activity	10^{-3}M, 10^{-4}, 10^{-5}M	Ahrabi et al. (2011), Esmaeili, et al. (2012)

Natural concentrations of free phenolic acids in soil under *C. canadensis* were up to 10^{-5}M. According to data from literature (Table 6), inhibitory concentrations of phenolic compounds for some physiological process in plants (plant-water relationships, stomatal conductance, seed germination, root and shoot elongation, chlorophyll fluorescence, PSII photochemistry, non-photochemical fluorescence quenching, and heat energy dissipation, chlorophyll, porphyrin and chlorophyllide contents, and chlorophyllase a and b activities, carotenoids and anthocyanin content, guaicol peroxidase activity) range from 10^{-6} to 10^{-3}M (Ahrabi et al., 2011; Esmaeili et al., 2012; Rasmunsen and Einhelling, 1978; Hussain and Reigosa, 2011; Blum, 1996; Barkosky et al., 2000; Aliotta et al., 1993; Barkosky and Einhellig, 1993; Yang et al., 2002, 2004; Djurdjević et al., 2008). The synergism of several phenolic compounds in inhibition of seed germination and seedling growth is also significant (Einhelling and Ramussen, 1978). The accumulation of phenolic toxins in soil causes either the reduction of number of plant individuals or complete elimination of some herbaceous plants and tree seedlings. Due to the allelopathic activity, the phenols strongly affect the structure and floristic composition of plant communities (Chou and Muller, 1972; Lodhi, 1975; Wardle et al., 1998; Djurdjević et al., 2004).

After the phase of complete domination of *C. canadensis* and its negative allelopathic influence on the other species in the phytocoenosis, the later phases of vegetation succession are marked by a reduction in its abundance, in the amount of leached allelochemicals and in their inhibitory effect, which leads to an increase in the abundance of the other herbaceous species. Yoda et al. (1963) showed that density-dependant mortality of *C. canadensis* seedlings occurred only in populations of 1000 plants m^{-2} or more. Autotoxicity also seems to be involved in so-called successional series in other ways. Speeding up the replacement of one species by another through self-allelopathic effects, Keever (1950) has shown that in old-field succession in North Carolina, horseweeed rapidly loses dominance after the first year because of the toxic effects of its own decay products, combined with the competitive effects of other species. All this suggests that autotoxicity is involved in the second-year stunting of the species.

In addition, the dominant invasive species *C. canadensis* creates allelopathically active matter, above all phenolic compounds which are washed away from its aboveground parts through water precipitation, leached through its roots, or released through the microbial degradation of dead plant remains. Having reached the soil in large enough quantities, the phenols inhibit seed germination and seedling growth of the other species in the ruderal phytocoenosis. Because of this, herbaceous species are only found in small numbers or are completely absent from the ruderal phytocoenosis. Owing to the high density of the plants per unit area, the height of the stems, and the high production of organic mass (Yoda et al., 1963; Bruce and Kells, 1990; Weaver, 2001), horseweed plays the key role in changing the infertile, inert sand into soil, which can be categorised as young sandy soil in the first phases of pedogenesis. As a consequence of *C. canadensis* autotoxicity, there occurs an increase in the number of individuals of other herbaceous species and the transition to the next phase of succession of the ruderal phytocoenosis with meadow elements.

SEASONAL DYNAMICS OF TOTAL PHENOLICS AND PHENOLIC ACIDS IN THE TISSUE OF THE ABOVEGROUND PARTS OF *C. CANADENSIS*

Figure 3A presents the results of the seasonal measurements of the amount of total phenolics (free and bound forms) in the vegetative parts of *C. canadensis* from March, when the plants where in the rosette phase, to September, when the plants were in the phase of fruit abscission and the beginning of plant decline. The amount of total free phenolics gradually increased (from 1,898 to 4,973 $\mu g\ g^{-1}$) between the rosette phase and the phase of intensive plant growth (March-June). At the beginning of flowering and during the phase of full flowering and fruiting (July-August), there is a sudden and multiple increase in the amount of total free phenolics in comparison to the previous phases (up to 31,000 $\mu g\ g^{-1}$). In the *C. canadensis* stem, the amount of bound phenolics is several times lower than free forms and they are at their peak during the initial phase of elongation (May) and intensive plant growth (June), and also in September, during the phase of fruit abscission and the beginning of plant decline.

In the tissue of this species, we detected 5 phenolic acids in their free forms (*p*-coumaric, ferulic, *p*-hydroxybenzoic, vanillic and syringic), reaching their highest levels in May (the beginning of plant elongation). Ferulic and *p*-coumaric acids, which are derivatives of cinnamic acid, predominate, while *p*-hydroxybenzoic, vanillic and syringic acids (derivatives of benzoic acid) are present in lower quantities for the majority of the vegetation season (Figure 3B). Bound forms of phenolic acids (Figure 3C) peak twice, in May and August, when ferulic acid predominates (up to 952 $\mu g\ g^{-1}$).

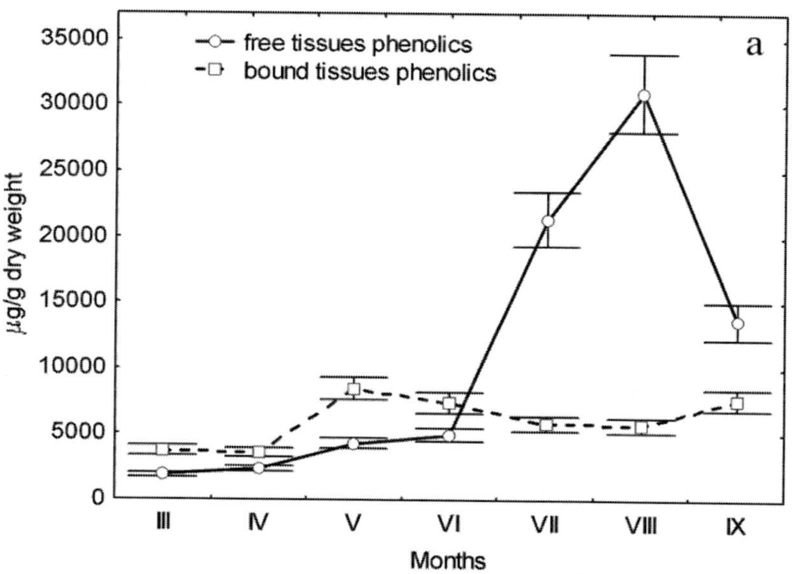

Figure 3. (Continued).

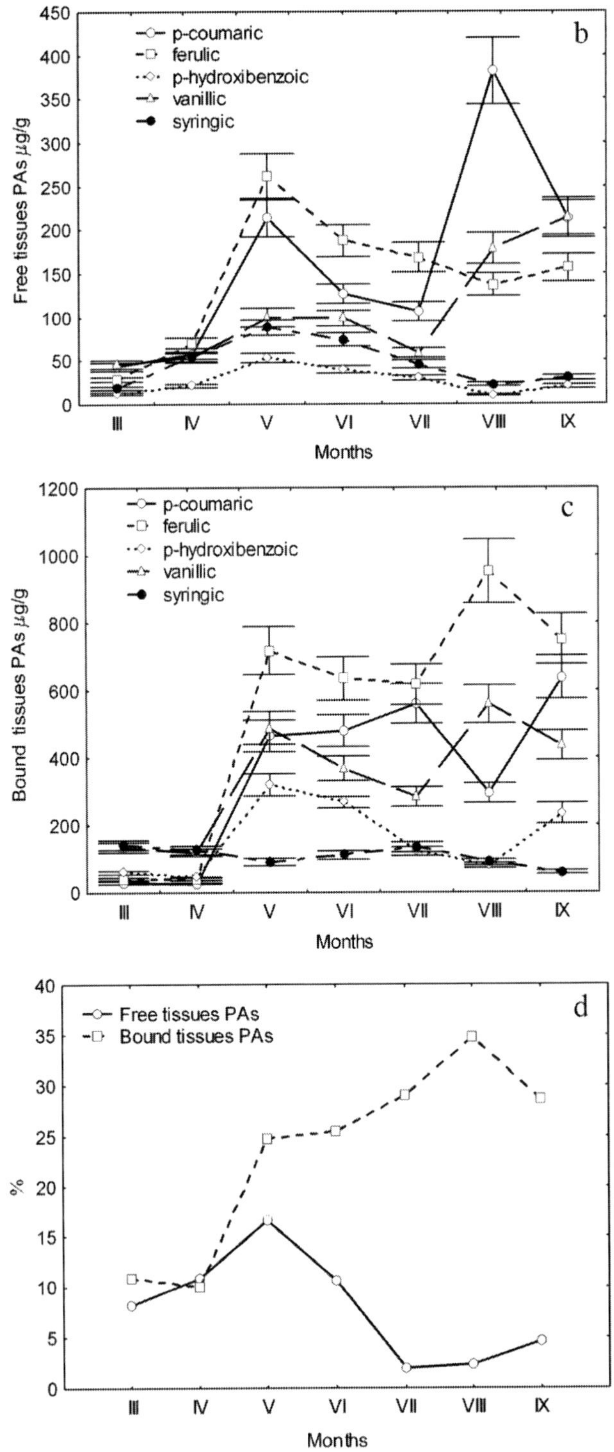

Figure 3. Seasonal dynamics of total phenolics and phenolic acids (free and bound forms) in *C. canadensis* plants (A-total phenolics; B-free phenolic acids; C-bound phenolic acids; D-percentage of phenolic acids in relation to the total phenols).

The amount of bound phenolic acids in the tissue is several times higher than free forms, particularly at the beginning of plant elongation (May) right up to the phase of fruit abscission and plant decline (September). If one considers the percentage ratio of free and bound forms of phenolic acids to the total free and bound phenolics during vegetation (Figure 3D), it can be seen that free acids reach their highest concentrations (16.7%) in May (the beginning of elongation), after which their amount decreases sharply until July (the beginning of flowering); thereafter it increases gradually to 4.6% by the end of vegetation. On the other hand, the percentage of bound phenolic acids increases markedly between April (young plants in the form of rosettes, 5-10cm in diameter) and August, during the phase of full flowering and the beginning of fruiting (34.7%). In summary, the trend of the seasonal dynamics of bound phenolic acids is opposite to that of free ones with their percentage in the aboveground parts of *C. canadensis* increasing up to the end of vegetation, while free acids decrease.

In terms of allelopathy, *C. canadensis* is a species that has not been studied enough either in native areas or introduced areas. As a pioneer weed species, it dominates during the first few years and is of great importance in vegetation succession in abandoned fields and in anthropogenically disturbed landscapes (Booth, 1941; Keever, 1950; Rice, 1974, 1979; Kapustka and Moleski, 1976; Shaukat et al., 2003; Gao et al., 2009; Djurdjević et al., 2011, 2012; Zaplata et al., 2011). As far as we know, this is the first work to study the seasonal dynamics of phenolic compounds which are of high allelopathic efficiency in the tissue of *C. canadensis*, during the plant's development phases and in the soil under it, in non-native areas. As a result of taking samples of plant material at monthly intervals, we were able to measure precisely the amount of total phenolics and phenolic acids in the aboveground parts of *C. canadensis* from the beginning of growth, when the plants were in rosette form, to fruit abscission and plant senescence. Our data on the seasonal dynamics of phenolics in the tissue of *C. canadensis* represent a step forward in the study of natural concentrations of phenolics as important allelochemicals of this in Europe and many other regions of the world alien species.

According to literature surveys, allelochemicals in plants have usually been analysed and measured only annually or three times a year in line with the seasons: spring, summer and autumn (e.g., Rice, 1974; Rice and Pancholy, 1973, 1974; González et al., 1995; Hyder et al., 2002a, b; Gallet and Lebreton, 1995; Gallet and Pellissier, 1997; Gallet et al., 1999; Reigosa et al., 1999a; Djurdjević et al., 1997, 1998, 2008; Wardle et al., 1998). There are copious data which support the novel weapon hypothesis for the dominance of invasive plants, according to which many plant invaders interfere with native plants through allelopathy. Phenolic compounds [(±)-catechin, apigenin and luteolin, chlorogenic gallic, caffeic, vanillic, *p*-coumaric, syringic and ferulic acids] synthesized in the tissue of an invasive species play a main role in this process (Cipollini et al., 2008; Dorning and Cipollini, 2006, Batish et al., 2006; Thorpe et al., 2009; He et al., 2009; Hierro and Callaway, 2003; Callaway and Ridenour, 2004; Callaway and Maron, 2006). In the tissue of *C. canadensis* we found 5 phenolic acids in their free forms (*p*-coumaric, ferulic, *p*-hydroxybenzoic, vanillic and syringic), with highest concentrations in May (the beginning of plant elongation). Ferulic and *p*-coumaric acids, which are derivatives of cinnamic acid, predominated, while *p*-hydroxybenzoic, vanillic and syringic acid (derivatives of benzoic acid) were present in lower amounts for the majority of the vegetation season. The amount of free phenolic acids was several times lower in the tissue of *C. canadensis* than bound forms. Free phenolic acids can have a direct allelopathic effect on surrounding plants after having been released into sandy soil, while bound forms can only do so after the microbial degradation of lignin from dead plant parts.

In the tissue of *C. canadensis*, up to 31 mg g^{-1} of total free phenolics were measured during the plant's elongation phase and the phase of fruit abscission and plant senescence, and 5 phenolic acids, which accounted for 7.4% of the total phenolics. According to data from literature, the following phenolics are found in the stems: gallic acid, vanillic acid, syringic acid and catechol (Shaukat et al., 2003), eugenyl β-Psd, scutellarin, luteolin-7-O-β-D-glucuronide, quercetin, quercetin-3-O-β-D-glucopyranosid and luteolin (Liu Hong-Li et al., 2011). Inflorescences of *C. canadensis* contain many more polyphenols, flavonoids (as apigenin) and some organic acids than leaves and shoots. Condensed polyphenols predominate in shoots (Vinogradova, 2007). We suppose that in our case, due to the higher amount of phenolics in *C. canadensis* inflorescences than in leaves and stems, there is an increase in the amount of free phenolics at the beginning of flowering and in the phase of full flowering and beginning of fruiting (21.3 and 31 mg g^{-1}, respectively) during July and August.

The trend of the seasonal dynamics of bound phenolic acids is opposite to that of free ones as their percentage in the aboveground parts of *C. canadensis* increased by the end of the vegetation season, while free acids decreased, which occurred due to incorporation into the structure of lignin, the most abundant polymer of woody and also slightly lignified herbaceous plants, like *C. canadensis*. Ferulic and *p*-coumaric acids are the main phenolic acids present in the cell wall of monocotyledons and especially of Poaceae. It seems likely that dimerization reactions occur in the cell walls when graminaceous plants are growing. Such a mechanism could be involved in cross-linking adjacent *p*-coumaroyl and feruloyl groups in arabinoxylans, leading to increased cell wall rigidity and decreased biodegradability of the wall polysaccharides (Hartley et al., 1990). Theodoru et al. (1987) showed that cell walls of barley stems contain *p*-coumaric (6.8 mg g^{-1}), ferulic (3.4 mg g^{-1}) and *p*-hydroxybenzoic acid (0.1 mg g^{-1}). These acids become incorporated into cell wall lignin, as demonstrated by Lowry et al. (1993). Phenolic constituents (lignins and phenolic acids) and carbohydrates are assembled in a tight architecture, which differs according to the plant species. Ferulic and diferulic acid are esterified with the polysaccharides of the cell walls (Scalbert et al., 1985; Kamisaka et al., 1990; Lewis and Yamamoto, 1990; Besle et al., 1994; Kondo et al., 1994; Provan et al., 1994). Information about these and general biosynthetic pathways in the plant's metabolism of phenolic substances is provided by Hättenschwiler and Vitousek (2000).

Seasonal Dynamics of Total Phenolics and Phenolic Acids in the Sandy Soil of a Ruderal Meadow Community Dominated by *C. canadensis*

The results of the seasonal measurements of the amount of total phenolics (free and bound) in the soil under *C. canadensis* are given in Figure 4A. Bound phenolics can be seen to have clear seasonal dynamics. Their amount falls rapidly from March (179 μg g^{-1}), when the plants are in the rosette phase, to June (109 μg g^{-1}), when the levels begin to increase again. On the other hand, changes in the amount of free phenolics are not so marked and they are at their lowest levels from May to July (the beginning of elongation and the beginning of the flowering of plants). In the soil, the amount of bound phenolics is several times higher than free forms, which is the case for the whole period of vegetation.

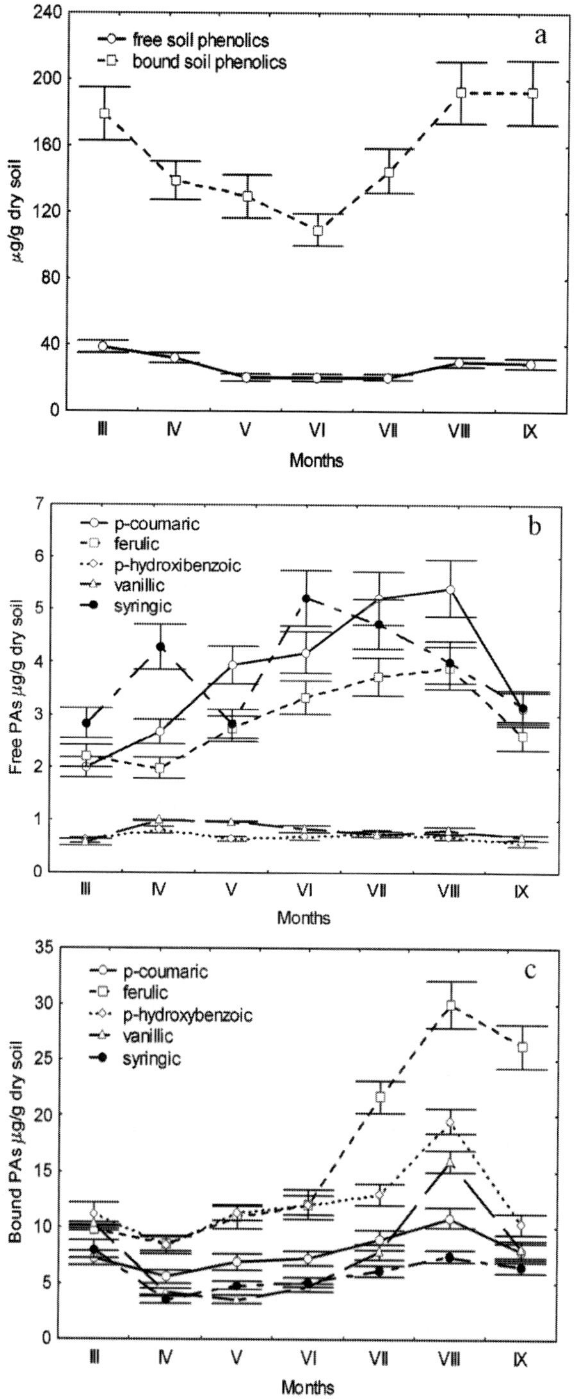

Figure 4. (Continued).

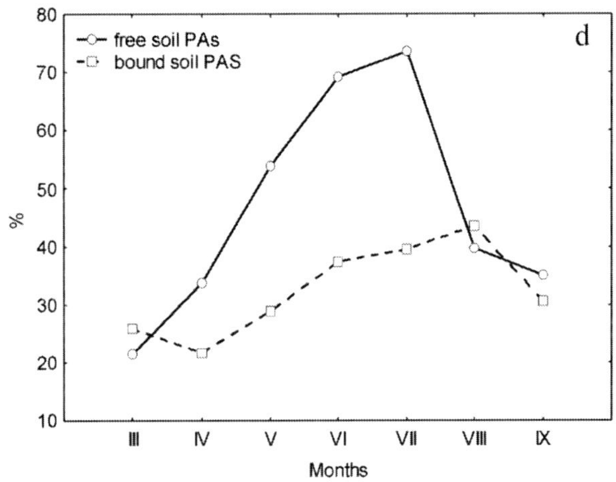

Figure 4. Seasonal dynamics of total phenolics and phenolic acids (free and bound forms) in sandy soil under *C. canadensis* (A-total phenolics; B-free phenolic acids; C-bound phenolic acids; D- percentage of phenolic acids in relation to the total phenols).

Only three free forms of phenolic acids in the soil under *C. canadensis* have clear seasonal dynamics (p-coumaric, ferulic and syringic acids), i.e., their amounts increase from March, when the plants are in the rosette phase, to August or the phase of full flowering and the beginning of fruiting (Figure 4B). Their amounts vary from 2-2.83 $\mu g\ g^{-1}$ at the beginning of vegetation up to 5.24 $\mu g\ g^{-1}$ during August. The amounts of p-hydroxybenzoic and vanillic acid do not change significantly during the period of vegetation and range from 0.57 to 0.99 $\mu g\ g^{-1}$. The amount of bound phenolic acids in the soil falls insignificantly between March and April, after which there is a gradual rise until July (the beginning of flowering), with all five phenolic acids reaching their highest levels in August during the phase of full flowering and the beginning of fruiting (Figure 4C). Ferulic (30.0 $\mu g\ g^{-1}$), p-hydroxybenzoic (19.56 $\mu g\ g^{-1}$) and vanillic acids (15.92 $\mu g\ g^{-1}$) predominate.

The percentage of free and bound phenolic acids in terms of the total free and bound forms of phenolics in the soil under *C. canadensis* during the vegetative period is shown in Figure 4D. It can be seen that the level of free acids as a percentage of the total free phenolics increases dramatically between March, when the plants are in the form of rosettes, 1-2 cm in diameter, and July (the beginning of flowering), when free phenolic acids account for as much as 73.7% of the total free phenolics in the soil. The percentage of bound phenolic acids gradually increases between April, when the plants are in the form of rosettes, 5-10 cm in diameter, and August, when they reach their maximum level of 43.5%. During the phase of fruit abscission and the beginning of plant senescence (September), the percentages of free and bound phenolic acids in the soil are almost equal (35.1% and 30.6%, respectively).

Correlation between Total Phenolics and Phenolic Acids in the Tissue of *C. canadensis* and the Sandy Soil

Data on the correlation between total phenolics (free and bound forms) and phenolic acids (free and bound forms) in the stem of *C. canadensis* and the soil under it is given in

Table 7. There is a positive correlation between total free phenolics in the plant tissue and the total bound phenolics and bound phenolic acids in the soil ($p < 0.01$). There is also a positive correlation between the total bound phenolics in the tissue of *C. canadensis*, on the one hand, and free and bound phenolic acids in the tissue of this species, on the other ($p < 0.05$). There is a significant negative correlation between total bound phenolics in the tissue of *C. canadensis* and total free phenolics in the soil ($p < 0.001$). The same is true for the correlation between the amount of bound phenolic acids in the tissue of the plant and the total free phenolics in the soil. As the amount of total free phenolics in the soil increases, so too does the amount of bound forms in it ($p < 0.001$). There is also a positive correlation between free phenolic acids in the plant, on the one hand, and bound phenolic acids in the plant and free phenolic acids in the soil, on the other ($p < 0.05$).

Table 7. Correlation coefficients between total phenolics and phenolic acids in *C. canadensis* tissues and associated sandy soil

	CFPh	CBPh	SFPh	SBPh	CFPAs	CBPAs	SFPAs	SBPAs
CFPh	1.00	0.10	-0.15	*0.50***	0.25	0.15	0.21	*0.51***
CBPh		1.00	*-0.75****	-0.25	*0.41**	*0.41**	0.85	0.04
SFPh			1.00	*0.65****	-0.28	*-0.33**	-0.21	0.05
SBPh				1.00	0.35	-0.05	-0.09	*0.39**
CFPAs					1.00	*0.34**	*0.38**	0.19
CBPAs						1.00	-0.02	-0.26
SFPAs							1.00	-0.27
SBPAs								1.00

*, **, *** Significantly different at 0.05, 0.01 and 0.001 level, respectively, by t-test (n = 5). CFPh, *C. canadensis* free total phenolics; CBPh, *C. canadensis* bound total phenolics; SFPh, soil free total phenolics, SBPh, soil bound total phenolics; CFPAs, *C. canadensis* free phenolic acids; CBPAs, *C. canadensis* bound phenolic acids; SFPAs, soil free phenolic acids; SBPAs, soil bound phenolic acids.

The accumulation of total bound phenolics in the soil under *C. canadensis* occurs between June, when the plants are in the phase of intensive growth (up to 1.5 m in height), and August, when the plants are in the phase of full flowering and fruiting. At the same time, in the period from June to August, plant tissues contain the highest concentration of total free phenols which can be easily released into the soil by precipitation during the vegetation period and after plant decay due to the only low lignification of *Conyza* cell walls. However, the amount of free and bound phenolic acids as a percentage of the total free and bound phenolics in the soil increases from the first rosette phases in plant development (March-April) right up to the phase of full flowering and fruiting (July-August). During that period, the release of allelochemicals into the soil occurs via leachates and volatiles from live or dead plant roots, leaves, stems, branches and the inflorescence, which coincides with the findings of many authors on many other dominant species in such phytocoenoses (Koeppe et al., 1976; Tang and Young, 1982; Rice, 1974; Inderjit and Dakshini, 1994, 1995; Inderjit, 1996; Batish et al., 2009; Lorenzo et al., 2011). Roots exude a variety of low-molecular weight organic compounds. Some of them, especially the phenolics (*p*-hydroxybenzoic, caffeic, *p*-coumaric, ferulic, gallic, gentisic, protocatechuic, salicylic, sinapic, syringic, catechin), influence the growth and development of surrounding plants and soil microorganisms (Bertin et al., 2003; Thorpe et al., 2009; Batish et al., 2006, 2009), in addition to influences upon organismic interactions (directing by allelopathy the competition between plants, supporting beneficial symbioses, influencing resistances against pests, etc.).

The whole *C. canadensis* plant (stem, branches and leaves) is covered with hairs and glandular trichomes (Weaver, 2001). They are sources for the release of phenolic compounds, which are secreted and washed away by precipitation into the soil under the plant. *C. canadensis* is characterised in its growing places by a large number of individuals (in May, 1000 young plants, 10–15 cm in height, per m^2 were counted) and a large aboveground mass, as the plants reach a height of up to 2 m in summer. Hence, phenolics released via the surface of the plant account for a significant proportion of their total amount. The release of phenolic compounds via the hairs and glandular trichomes of various plant species has been documented by the research of many authors (Avé and Tingey, 1986; Sterling et al., 1987; Wagner, 1991; Salminen et al., 2004).

The investigated soil at the levee in the pioneer meadow community can be categorised as young sandy soil in the first phases of pedogenesis, in which the dominant species *C. canadensis* has an important role to play, due to the decomposition of its dead organic remains and/or the compounds released during vegetation by its roots and aboveground parts. Although the dominant *C. canadensis* creates high levels of free phenols (31 mg g^{-1}), only an insignificant percentage (0.6%) are bound to sand particles because of the low absorbent capacity of the latter, as compared to soil particles. As consequence, quite an amount of the phenolics released from the plants are washed away by precipitation into deeper layers and hence become inaccessible to plants. The significantly higher percentage of free phenolic acids for the majority of the growing season (maximum 73.7% in July) when compared to bound forms (maximum 43.5% in August) highlights this, too. Their presence in the sandy soil under *C. canadensis*, although in smaller amounts than in natural soils, indicates the initiation of pedogenetic processes, processes of organic matter transformation, and humus formation. On the other hand, the presence of phenolics as secondary plant metabolites points to the influence of these invasive dominant plants in the overbuilding of substrate, such as sand deposits (Djurdjević et al., 2011). Gordon (1998) hypothesizes that many invasive non-indigenous species do alter ecosystem properties at several levels. Out of 31 species studied in Florida, 12-20 (39-64%) potentially alter ecosystem properties of geomorphology, hydrology, biogeochemistry, and disturbance. Also, Weidenhamer and Callaway (2010) emphasize that invasive species can alter the biogeochemistry of ecosystems, as secondary metabolites released by invasive species may play important roles in soil chemistry, as well as plant-plant and plant-microbe interactions.

As opposed to our monitoring of the dynamics of phenolics in soil at monthly intervals during the vegetation period of the dominant invasive *C. canadensis*, data from other such cases are available only seasonally (spring, summer, autumn) and not directly related to the plant ontogenetic stages. Lodhi (1975) isolated ferulic, caffeic, and *p*-coumaric acids from soils under hackberry (*Celtis laevigata*) trees. There, the phytotoxicity of each phenolic acid was highest in January and April and lowest in September. However, phenolic concentrations in soils collected in August were not as high as those collected in January and April. The highest concentration of caffeic acid in the humus of bilberry (*Empetrum hermaphroditum*) in northern Sweden was recovered during snowmelt in May (Gallet and Lebreton, 1995). Muscolo and Sidari (2006) examined seasonal variations in the content of water-soluble phenolics extracted from litter and the Ah horizon of forest soil under *Pinus laricio* Poiret trees. Similar to our findings, the results indicated a seasonal variation of phenolic substances contents in soil, with the highest concentrations of water-soluble phenolic acids in autumn and the lowest in summer.

According to our previous research, the seasonal dynamics of phenolics in soil depends on the type of phytocoenoses and on their dominant species. Among forests in Serbia, in a sessile oak forest the highest amount of total phenolics occurs in summer (Djurdjević and Oberan, 1998), the amount falls between spring and autumn in a Scots pine and European black pine forest (Djurdjević et al., 1997), it falls in spring, but increases during summer and autumn in a forest of black ash and Virgilius oak (Djurdjević et al., 1998), and it falls between March and May in the rhizosphere soil of the vernal ephemeroid *Arum maculatum* in a beech and linden forest (Djurdjević et al., 2008).

Kuiters and Denneman (1987) studied monthly fluctuations in soil phenolics content (free and bound forms) under stands with *Fagus sylvatica*, on strongly podzolized sandy soil. A distinct seasonal pattern could be found in the soil phenolic content at this site. The $Ca(OH)_2$-extractable concentration, including water-soluble as well as reversibly bound phenolics, rose to 139% during and after the leaf-fall period and reached the highest concentration in late winter (January). Thereafter, from January to March, the concentration fell to its former level and was stable until leaf-fall in autumn. The water–soluble concentration showed a seasonal pattern as well, but the extent of these fluctuations was much smaller. The authors argued that the fall in phenolic acids in spring was the result of increased microbial activity due to increased soil temperatures and to the soil's adsorption of organic matter during humification. In the soil under the invasive *C. canadensis* analyzed in present investigation, the concentration of free phenolics also changed insignificantly during the growing period, as opposed to bound phenolics, where the changes were more marked and significant. The decrease in the amount of bound phenolics between March and June can probably also be attributed to increased microbial activity due to high soil moisture and temperature in early summer at this open ruderal meadow community on a sandy levee, and to the soil's adsorption of organic matter during humification. A substantial amount of phenolic compounds may have been transported also into the deeper layers of the sandy soil, which is not able to absorb them due to the coarseness of its particles. Some of the phenols are further degraded also by many organisms which used them for energy gain and syntheses in cell metabolism (Felbeck, 1971; Haworth, 1971; Martin and Haider, 1971).

The phenolic compounds originating from *C. canadensis* tissue, in addition to the phenolics which occur through microbial degradation of organic material, mainly lignin, and those phenolics which are produced by soil microorganisms, enter the soil and become part of the humus fraction. Humic acid, the most abundant constituent of humus, and the similar type fulvic acid, is a complex polymer of various phenolic units with linked peptides, amino acids and some polysaccharides (Martin and Haider, 1969, 1971; Haider et al., 1975; Flaig, 1971; Felbeck, 1971; Haworth, 1971). Phenolic compounds make up an important component of the humus layer of different soil types (Guenzi and McCalla, 1966; Hennequin and Juste, 1967; Katase, 1981a, b; Kögel, 1986; Kuiters and Denneman, 1987; Djurdjević et al., 2000, 2003, 2004, 2005a, 2006, 2009, 2012; Banach-Szott, 2007).

Conclusion

The domination of *C. canadensis* as the globally successful invader is achieved through the production of secondary metabolites, primarily phenolic compounds. Phenolic compounds accumulated in the soil under *C. canadensis* had an inhibitory effect on seed germination and seedling growth of surrounding herbaceous plants, reducing their abundance and cover.

The greatest amount of phenolic compounds was synthesized in the tissues during the most intense phase of plant growth, flowering and fruiting (July-beginning of flowering, and August-full flower and beginning of fruiting, plants were up to 2 m in height).

The level of free acids in the soil increased dramatically between March, when the plants were in the form of rosettes, 1-2 cm in diameter, and July (the beginning of flowering), when free phenolic acids account for as much as 73. 7% of the total free phenolics in the soil.

The globally successful invader *C. canadensis* plays a decisive role in the first phases of vegetation succession and the process of soil formation on the barren sandy levees, owing to the synthesis of secondary phenolic metabolites.

Acknowledgements

This work was supported by the Ministry of Education, Science and Tehnological Development of Serbia, grant No.173018. The authors would like to thank Dr. Anka Dinić, a retired Senior Scientist, for her extensive help during field research into the domination of *C. canadensis* and vegetation succession on sandy levees. We thank Jonathan Pendlebury, lector at Belgrade University's Faculty of Philology, for proofreading this paper.

References

Ahrabi, F., Enteshari, S., Moradshahi A. (2011). Allelopathic potential of para-hydroxybenzoic acid and coumarin on canola: Talaieh cultivar. *Journal of Medicinal Plant Research*, 5(20), 5104-5109.

Alexander, M. *Introduction to soil microbiology*. New York. J. Wiley and Soons. (1977).

Aliotta, G., Cafiero, G., Fiorentino, A., Strumia, S. (1993) Inhibition of radish germination and root growth by coumarin and phenylpropanoids. *Journal of Chemical Ecology*, 19, 175-183.

Avé, D. A, Tingey, W. M. (1986). Phenolic constituents of glandular trichomes on *Solanum berthaultii* and *S. polyadenium*. *American Journal of Potato Research*, 63, 9473-9480.

Banach-Szot, M. (2007). Content of phenolic compounds in humus acids of forest soil. *Polish Journal of Soil Science,* 40, 67-77.

Barkosky, R. R., Einhellig, F. A. (1993). Effects of salycilic acid on plant-water relationships. *Journal of Chemical Ecology*, 19, 237-247.

Barkosky, R. R., Einhellig, F. A., Butler, J. L. (2000). Caffeic acid-induced changes in plant–water relationships and photosynthesis in leafy spurge *Euphorbia esula*. *Journal of Chemical Ecology*, 26, 2095-2109.

Barto, E. K., Powell, J. R., Cipollini, D. (2010). How novel are the chenical weapons of garlic mustard in North American forest understories? *Biologica Invasions*, 12, 3465-3471.

Bate-Smith, E. C. *Flavonoid patterns in the Monocotyledons.* In: Perspectives in phytochemistry. Eds. Harborne, J. B. and Swain, T. Acad. Press London-New York. (1969). pp.167-177.

Batish, D. R., Kaur, S., Singh, H. P., Kohli, R. K. (2009). Role of root-mediated interactions in phytotoxic interference of *Ageratum conyzoides* with rice (*Oryza sativa*). *Flora* 204, 388-395.

Batish, D. R., Singh, H. P., Rana, N., Kohli, R. K. (2006). Assessment of allelopathic interference of *Chenopodium album* through its leachates, debris extracts, rhizosphere and amended soil. *Archives of Agronomy and Soil Science*, 52, 705-715.

Bertin, C., Yang, X., Weston, L. A. (2003). The role of root exudates and allelochemicals in the rhizosphere. *Plant and Soil*, 256, 67-83.

Besle, J. M., Cornu, A., Jouany, J. P. (1994). Roles of structural phenylpropanoids in forage cell wall digestion. *Journal of the Science of Food and Agriculture*, 64, 171-190.

Bhowmik, P. C., Bekech, M. M., (1993). Horseweed (*Conyza canadensis*) seed production, emergence, and distribution in notillage and conventional-tillage corn (*Zea mays*). *Agronomy (Trends in Agric. Sci.)*, 1, 67-71.

Blum, U. (1996). Allelopathic Interactions Involving Phenolic Acids. *Journal of Nematology*, 28, 129-132.

Booth, W. E., (1941). Revegetation of abandoned fields in Kansas and Oklahoma, *American Journal of Botany,* 28, 415-422.

Bousquet-Mélou, A., Louis, S., Robles, C., Greff, S., Dupouyet, S., Fernandez, C. (2005). Allelopathic potential of *Medicago arborea*, a Mediterranean invasive shrub. *Chemoecology*, 15, 193-198.

Bruce, J. A., Kells, J. J. (1990). Horseweed (*Conyza canadensis*) control in no-tillage soybeans (*Glycine max*) with preplant and preemergence herbicides. *Weed Technology*, 4, 642-647.

Callaway, R. M., Maron, J. L. (2006). What have exotic plant invasionstaught us over the past 20 years? *Trends in Ecology and Evolution*, 21, 369-374.

Callaway, R. M., Ridenour, W. M. (2004). Novel weapons: invasive success and the evolution of increased competitive ability. *Frontiers in Ecology and Environment,* 2, 436-443.

Carballeira, A., Reigosa, M. J. (1999). Effects of natural leachates of *Acacia dealbata* Link in Galicia (NW Spain). *Botanical Bulletin Academia Sinica*, 40, 87-92.

Chou, C. H., Lee, Y. F., (1991). Allelopathic dominance of *Miscanthus transmorrisonensis* in an alpine grassland community in Taiwan. *Journal of Chemical Ecology,* 17, 2267-2281.

Chou, C. H., Muller, C. H., (1972). Allelopathic mechanisms of *Arctostaphylos glandulosa* var. *zacaensis*. *American Midland Naturalist*, 88, 324-347.

Cipollini, D., Stevenson, R., Enright, S., Eyles, A., Bonello, P. (2008). *Phenolic metabolites in leaves of the invasive shrub, Lonicera maackii, and their potential phytotoxic and anti-herbivore effects.* *Journal of Chemical Ecology*, 34, 144-152.

Dinić, A., Stojšić, V., Pavlović, P., Djurdjević L., Mitrović, M., Kostić, O. Olđa, M., Vukadinović, B., (2005). *Monitoring of regeneration of sands vegetation after foresm fire in 1996 on the Deliblato Sands.* The Deliblato Sands special nature reserve Proceedings VII. 299-312.

Djurdjević, L. (1991). Allelopathic effects of blackberry (*Rubus hirtus* W. K.) *Archives of Biological Sciences* (Belgrade), 34, 31-38.

Djurdjević, L., Dinić, A., Kuzmanović, A., Kalinić, M. (1998). Phenolic acids and total phenols in soil, litter and dominating plant species in community *Orno-Quercetum virgilianae* Gajić 1952. *Archives of Biological Sciences,* (Belgrade), 50, 21-28.

Djurdjević, L., Dinić, A., Mitrović, M., Pavlović, P., Tešević, V. (2003). Phenolic acids distribution in a peat of the relict community with Serbian spruce in the Tara Mt. Forest reserve (Serbia). *European Journal of Soil Biology*, 39, 97-103.

Djurdjević, L., Dinić, A., Pavlović, P., Mitrović, M., Karadžić, B., Tešević, V. (2004). Allelopathic potential of *Allium ursinum* L. *Biochemical Systematics and Ecology,* 32, 533-544.

Djurdjević, L., Dinić, A., Stojšić, V., Mitrović, M., Pavlović, P., Oldja, M. (2000). Allelopathy of *Paeonia officinalis* L. 1753 *ssp. banatica* (Rochel) Soó 1945, a Pannonian endemic and relict species. *Archives of Biological Sciences* (Belgrade), 52, 197-203.

Djurdjević, L., Milenković, M., Pavlović, P., Kostić, O. (1999). Allelopathic investigations in the *Fraxino angustifoliae-quercetum roboris* (Jov. et Tomić, 1979) forest community with the autumnal truffle (*Tuber macrosporum* Vitt.). *Archives of Biological Sciences* (Belgrade), 51, 27-33.

Djurdjević, L., Mitrović, M., Dinić, A., Pavlović, P., Bojović, S., Gajić, G., Kostić, O. (2005a). Allelopathic investigation of *Quercus conferta* and *Quercus cerris* domination in oak forest at Avala Mt. (Serbia). *Proceedings of the Fourth World Congress on Allelopathy,* 21-26 August 2005, Charles Sturt University, Wagga Wagga, Australia. Web site www.regional.org.au/au/allelopathy.

Djurdjević, L., Mitrović, M., Gajić, G., Jarić, S., Kostić, O., Oberan, L., Pavlović, P. (2011). An allelopathic investigation of the domination of the introduced invasive *Conyza canadensis* L. *Flora*, 206, 921-927.

Djurdjević, L., Mitrović, M., Pavlović, P., Gajić, G., Kostić, O., (2006). Phenolic acids as bioindicators of fly ash deposit revegetation. *Archives of Environmental Contamination and Toxicology,* 50, 488-495.

Djurdjević, L., Mitrović, M., Pavlović, P., Perišić, S., Mačukanović-Jocić, M. (2005b). Total phenolics and phenolic acids content in low (*Chrysopogon gryllus*) and medium quality (*Festucavallesiaca*) forage grasses of Deliblato Sands meadow-pasture communities in Serbia. *Czech Journal of Animal Science*, 50, 54-59.

Djurdjević, L., Oberan, V. Lj. (1998). Seasonal dynamics of phenolics and microorganisms in the litter of oak forest on Cer Mt. *Archives of Biological Sciences*(Belgrade), 50, 231-236.

Djurdjević, L., Pavlović, P., Mitrović, M. *The effect of phenolic compounds* on *soil properties.* In: Muscolo A, Sidari M (Eds) Soil Phenols. Nova Science Publishers, Inc. New York, (2009). pp 31-62.

Djurdjević, L., Popović, Z., Mitrović, M., Pavlović, P., Jarić, S., Oberan, L., Gajić, G. (2008). Dynamics of bioavailable rhizosphere soil phenolics and photosynthesis of *Arum maculatum* L. in a lime-beech forest. *Flora*, 203, 590-601.

Djurdjević, L., Dinić, A., Mitrović, M., Pavlović, P., *Allelopathy of Erica carnea, a dominant species in the mixed pine forests at Maljen mountain* (Serbia). In: Progress in Botanical Research. Tsekos J., Moustakas, M. (Eds.), Kluwer Academic Publishers. Dodrecht, Boston, London. (1997). pp. 289-293.

Djurdjević, L., Gajić, G., Kostić, O., Jarić, S., Pavlović, M., Mitrović, M. and Pavlović P. (2012). Seasonal dynamics of allelopathically significant phenolic compounds in globally successful invader *Conyza canadensis* L. plants and associated sandy soil. *Flora* 207, 812-820.

Dorning, M., Cipollini, D. (2006). Leaf and root extracts of the invasive shrub, *Lonicera maackii*, inhibit seed germination of three herbs with no autotoxic effects. *Plant Ecology*, 184, 287-296.

Einhelling, F. A., Rasmussen, J. A. (1978). Synergistic inhibitory effects of vanillic and *p*-hydroxybenzoic acids on radish and grain sorghum. *Journal of Chemical Ecology*, 4, 425-436.

Esmaeili, M., Heidarzade, A., Pirdashti, H. (2012). Inhibitory activity of pure allelochemicals on Barnyardgrass (*Echinochloa crus-galli* L) seed and seedling parameters. *International Journal of Agriculture and Crop Sciences* 4-6, 274-279.

Felbeck, G. T. J. R. (1971) Structural hypothesis of soil humic acids. *Soil Science*, 111, 42-48.

Flaig, W. (1971). Organic compounds in soil. *Soil Science*, 111, 19-33.

Gajić, M. Erigeron L. In: Josifović (ed) *Flora* SR Srbije, VII. (1975). pp 26-30.

Gallet, C., Lebreton, P. (1995). Evolution of phenolic patterns in plants and associated litters and humus of mountain forest ecosystem. *Soil Biology and Biochemistry*, 27, 157-165.

Gallet, C., Nilsson, M. C., Zackrisson, O. (1999). Phenolic metabolites of ecological significance in *Empetrum hermaphroditum* leaves and associated humus. *Plant and Soil*, 210, 1-9.

Gallet, C., Pellissier, F. (1997). Phenolic compounds in natural solutions of a coniferous forest. *Journal of Chemical Ecology*, 23, 2401-2412.

Gao, X., Li, M., Gao, Z., Zhang, H., Sun Z. (2009). Allelopathic effects of *Conyza canadensis* on the germination and growth of wheat, sorghum, cucumber, rape and radish. *Allelopathy Journal*, 23, 287-296.

González, L., Souto, X. C., Reigosa, M. J. (1995). Allelopathic effects of *Acacia melanoxylon* R. Br. phyllodes during their decomposition. *Forest Ecology and Management*, 77, 53-63.

Gordon, D. R. (1998). Effects of Invasive, Non-Indigenous Plant Species on Ecosystem Processes: Lessons from Florida. *Ecological Application*, 8, 975-989.

Grodzinsky, A. M., Golovko, E. A., Gorobec, S. A., Kostroma, E. J., Krupa, L. I., Prutenskaya, N. I., Seredjuk, L. S., Yurczak, L. D., Šrolj, T. S. (1987). *Experimental allelopathy.* Naukova dumka, Kiev (in Russian).

Guenzi, W. D., McCalla, T. M. (1966). Phytotoxic substances extracted from soil. *Soil Science Society of America, Proceedings,* 30, 214-216.

Haider, K., Martin, J. P., Filip, Z. *Humus biochemistry.* In: Paul E. A., McLaren A. D. (eds) Soil Biochemistry Vol. 4, Chap. 6. Marcel Dekker, New York, (1975). pp 195-244.

Harborne, J. B. (1980). *Plant phenolics.* In: Bell EA Charlwod BV (eds) Secondary plant products. Springer-Verlag, Berlin, Heidelberg, New York, Encyclopaedia of plant physiology, New series, 8, 329-402.

Hartley, R. D., Morrisson, W. H., Hilmmelsbach, D. S., Borneman, W. S. (1990). Cross-linking of cell wall phenolic arabinoxylans in graminaceous plants. *Phytochemistry*, 29, 3705-3709.

Hättenschwiler, S., Vitousek, P. M. (2000). The role of polyphenols in terrestrial ecosystem nutrient cycling. *Tree*, 15, 238-243.

Haworth, R. D. (1971). The chemical nature of humic acid. *Soil Science*, 111, 71-79.

He, W. M., Feng, Y., Ridenour, W. M., Thelen, G. C., Pollock, J. L., Diaconu, A., Callaway, R. M. (2009). Novel weapons and invasion: biogeographic differences in the competitive effects of *Centaurea maculosa* and its root exudate (±)-catechin. *Oecologia*, 159, 803-815.

Hennequin, J. R., Juste, C. (1967). Présence d'acides phenols libre dans le sol: Etude de leur influence sur la germination et la croissance des vegétaux. *Annual agronomy*, 18, 545-569.

Hierro, J. L., Callaway, R. M. (2003). Allelopathy and exotic plant invasion. *Plant and Soil*, 256, 29-39.

Hussain, M. I., González, L., Souto, C., Reigosa, M. J. (2011). Ecophysiological responses of three native herbs to phytotoxic potential of invasive *Acacia melanoxylon* R. Br. *Agroforestry Systems*, 149-166.

Hussain, M. I., Reigosa, M. J. (2011). Allelochemical stress inhibits growth, leaf water relations, PSII photochemistry, non-photochemical fluorescence quenching, and heat energy dissipation in three C_3 perennial species. *Journal of Experimental Botany*, 62, 4533-4545.

Hyder, P. W., Fredrickson, E. L., Estell, R. E., Lucero, M. E. (2002a). Transport of phenolic compounds from leaf surface of creosotebush and tarbush to soil surface by precipitation. *Journal of Chemical Ecology*, 28, 24775-2482.

Hyder, P. W., Fredrickson, E. L., Estell, R. E., Tellez, M., Gibbens, R. P. (2002b). Distribution and concentration of total phenolics, condensed tannins, and nordihydroguaiaretic acid (NDGA) in creosotebush (*Larrea tridentata*). *Biochemical Systematics and Ecology*, 30, 905-912.

Inderjit, (1996). Plant phenolics in allelopathy. *Botanical Review*, 62, 186-202.

Inderjit, Dakshini, K. M. M. (1994). Allelopathic effects of *Pluchea lanceolata* on characteristics of four soils and growth of mustard and tomato. *American Journal of Botany*, 81, 799-804.

Inderjit, Dakshini, K. M. M. (1995). On laboratory bioassays in allelopathy. *Botanical Review*, 61, 28-44.

Inderjit, Seastedt, T. R., Callaway, R. M., Pollock, J. L., Kaur, J. (2008). Allelopathy and plant invasions: traditional, congeneric, and bioeographical approaches. *Biological Invasions* 10, 875-890.

Jarić, S., (2009). *Allohtonous plant species in the natural and anthropogenically disturbed phytocoenoses of Srem.* Doctoral thesis, Faculty of Agronomy, University of Belgrade, 243 p. (in Serbian).

Jovanović, S., (1994). *Ecological study of ruderal flora and vegetation of Belgrade.* Doctoral thesis. Faculty of Biology, University of Belgrade, 222 p. (in Serbian).

Kamisaka, S., Takeda, S., Takahashi, K., Shibata, K. (1990). Diferulic and ferulic acid in the cell wall of Avena coleoptiles: Their relationships to mechanical properties of the cell wall. *Physiolgia Plantarum*, 78, 1-7.

Kapustka, L. A., Moleski, F. L. (1976). Changes in community structure in Oklahoma old field succession. *Botanical Gazette*, 137, 7-10.

Katase, T. (1981a) The different forms in which p-coumaric acid exists in a peat soil. *Soil Science*, 131, 271-275.

Katase, T. (1981b). The different forms in which p-hydroxybenzoic, vanillic, and ferulic acids exist in a peat soil. *Soil Science*, 132, 436-443.

Keever, O. (1950). Causes of succession on old fields of the Piedmont, North California, *Ecological Monographs*, 20, 229-250.

Kim, Y. O., Lee, E. J. (2011). Comparison of phenolic compounds and the effects of invasive and native species in East Asia: support for the novel weapons hypothesis. *Ecological Research*, 26, 87-94.

Koeppe, D. E., Southwick, L. M., Bittell, J. E. (1976). The relationship of tissue chlorogenic acid concentrations and leaching of phenolics from sunflowers grown under varying phosphate nutrient conditions. *Canadian Journal of Botany*, 54, 593-599.

Kögel, I. (1986). Estimation and decomposition pattern of the lignin component in forest humus layers. *Soil Biology and Biochemistry*, 18, 589-594.

Kögel-Knabner, I. (2002). The macromolecular organic composition of plant and microbial residues as inputs to soil organic matter. *Soil Biology and Biochemistry*, 34, 139-162.

Kondo, T., Ohshitsa, T., Kyuma, T. (1994). Comparison of phenolic acids in lignin fractions from forage grasses before and after digestion by sheep. *Animal Feed Science and Technology*, 47, 277-285.

Kuiters, A. T., Denneman, C. A. J. (1987). Water-soluble phenolic substances in soils under several coniferous and deciduous tree species. *Soil Biology and Biochemistry*, 19, 765-769.

Lewis, G. N., Yamamoto, E. (1990). Lignin: occurrence, biogenesis and biodegradation. Annu. Rev. Plant Physiol. *Plant Molecular Biology*, 41, 455-496.

Li, H. H., Nishimura, H., Hasegawa, K., Mixutani, J., (1992a). Allelopathy of *Sasa cernua*. *Journal of Chemical Ecology*, 18, 1785-1796.

Li, H. H., Urashima, M., Amano, M., Lajide, L., Nishimura, H., Hasegawa, K., Mizutani, J., (1992b). Allelopathy of Barnyardgrass (*Echinochloa crus-galli* L. Beauv. Var. *crus-galli*). *Weed Research* (Japan), 37, 146-152.

Liu Hong-Li, Liu Bai-Lian, Wang Guo-Cai, Dai Yi, YE Wen-Cai, LI Yao-Lan (2011). Studies on the chemical constituents from *Conyza canadensis*. *Journal of Chinese Medicinal Materials* 43(5). DOI：CNKI:SUN:ZYCA. 0.2011-05-023. ISSN: 1001-4454.

Lodhi, M. A. K. (1975). Soil-plant phytotoxicity and its possible significance in patterning of herbaceous vegetation in a bottomland forest. *American Journal of Botany*, 62, 618-622.

Lodhi, M. A. K., Rice, E. L. (1971). Allelopathic effects of *Celtis laevigata*. *Bulletin of the Torrey Botanical Club*, 98, 83-89.

Lorenzo, P., Palomera-Pérez, A., Reigosa, M. J., González, L. (2011). Allelopathic interference of invasive *Acacia dealbata* Link on the physiological parameters of native understory species. *Plant Ecology*, 212, 403-412.

Lowry, J. B., Sumpter E. A., McSweeney C. S., Schlink A. C., Bowden B., (1993). Phenolic acids in the fibre of some tropical grasses, effect on feed quality and their metabolism by sheep. *Australian Journal of Agricultural Research*, 44, 1123-1133.

Lyu, S. W., Blum, U. (1990). Effects of ferulic acid, and allelopathic compound on net P, K, and water uptake by cucumber seedlings in a split-root system. *Journal of Chemical Ecology*, 16, 2429-2439.

Martin, J. P., Haider, K. (1969). Phenolic polymers of *Stachybotrys atra, Stachybotrys chartarum* and *Epicoccum nigrum* in relation to humic acid formation. *Soil Science*, 107, 260-270.

Martin, J. P., Haider, K. (1971). Microbial activity in relation to soil humus formation. *Soil Science*, 111, 54-70.

McEwan, R. W., Arthur-Paratley L. G., Rieske L. K., Arthur, M. A. (2010). A multi-assay comparison of seed germination inhibition by *Lonicera maackii* and co-occurring native shrubs. *Flora*, 205, 475-483.

Mersie, W., Singh, M. (1988). Effects of Phenolic Acids and Ragweed Parthenium (*Parthenium hysterophorus*) Extracts on Tomato (*Lycopersicon esculentum*) Growth and Nutrient and Chlorophyll Content. *Weed Science*, 36, 278-281.

Muller, C. H., (1966). The role of chemical inhibition (allelopathy) in vegetation composition. *Bulletin of the Torrey Botanical Club*, 93, 332-351.

Murrell, C., Gerbe,E., Krebs, C., Parepa, M., Schaffner, U., Bossdorf, O. (2010). Invasive knotweed affects native plants through allelopathy. *American Journal of Botany*, 98, 38-43.

Muscolo, A., Sidari, M. (2006). Seasonal fluctuations in soil phenolics of a coniferous forest: effects on seed germination of different coniferous species. *Plant and Soil*, 284, 305-318.

Ni, G. Y., Schaffiner, U., Peng, S. L., Callaway, R. M. (2010). *Acroptilon repens*, an Asian invader, has stronger competitive effects on species from America than species from its native range. *Biological Invasions*, 12, 3653-3663.

Orr, S. P., Rudgers, J. A., Clay, K. (2005). Invasive plants can inhibit native tree seedlings: testing potential allelopathic mechanisms. *Plant Ecology*, 181, 153-165.

Pančić, J. (1874). *The flora of the Principality of Serbia: Vascular plants growing in the wild in Serbia.* State Publishing House, Belgrade (In Serian).

Pančić, J. (1878). *The flora in the Belgrade region: According to an analytical system.* State Publishing House, Belgrade (In Serbian).

Pisula, N. L., Meiners, S. J. (2010). Relative allelopathic potential of invasive plant species in a young disturbed woodland. *Journal of the Torrey Botanical Society,* 137, 81-87.

Provan, G. J., Scobbie, L., Chesson, A. (1994). Determination of phenolic acids in plant cell walls by microwave digestion. *Journal of the Science of Food and Agriculture*, 64, 63-65.

Ramussen, J. A., Einhelling, F. A. (1978). Synergistic inhibitory effects of *p*-coumaric and ferulic acids on germination and growth of grain sorghum. *Journal of Chemical Ecology*, 3, 197-205.

Regehr, D. L., Bazzaz, F. A., (1979). The population dynamics of *Erigeron canadensis*, a successional winter annual. *Journal of Ecology*, 67, 923-933.

Reigosa, M. J., Sánchez-Moreiras, A., González, L. (1999a). Ecophysiological approach in allelopathy. *Critical Reviews in Plant Sciences*, 18, 577-608.

Reigosa, M. J., Souto, X. C., Gonzalez, L. (1999b). Effect of phenolic compounds on the germination of six weed species. *Plant Growth Regulation*, 28, 83-88.

Rice, E. L. (1979). Allelopathy: An update. *Botanical Review*, 45, 15-109.

Rice, E. L., Pancholy, S. K. (1973). Inhibition of nitrification by climax ecosystems. II. Additional evidence and possible role of tannins. *American Journal of Botany*, 60, 691-702.

Rice, E. L., Pancholy, S. K. (1974). Inhibition of nitrification by climax ecosystems. III. Inhibitors other than tannins. *American Journal of Botany*, 61, 1095-1103.

Rice, E. L., (1974). *Allelopathy.* Academic Press New York, San Francisko, London.

Rice, E. L., (1979). Allelopathy-an update. *Botanical Review*, 45, 15-109.

Root, R. A., Wilson, R. E., (1973). Changes in biomass of six dominant plant species during old field succession in Southeastern Indiana. *Ohio Journal of Science*, 73, 370-375.

Salminen, J. P., Kapari, L., Lempa, K., Ossipov, V., Sinkkonen, J., Valkama, E., Haukioja, E., Pihlaja, K. (2004). Defensive effect of surface flavonoid glycones of *Betula pubescens* leaves against first instar *Epirrita autumnata* larvae. *Journal of Chemical Ecology*, 30, 2257-2268.

Saunders, J. A., McClure, J. W. (1976). The distribution of flavonoids in chloroplasts of 25 species of vascular plants. *Phytochemistry*, 15, 809-810.

Scalbert, A., Monties, B., Lallemand, J. Y., Guittet, E., Rolando, C. (1985). Ether linkage between phenolic acids and lignin fractions from wheat straw. *Phytochemistry*, 24, 1359-1362.

Séne, M., Gallet T., Doré T., (2001). Phenolic compounds in a Sahelian sorghum (*Sorghum bicolor*) genotype (CE145-66) and associated soils. *Journal of Chemical Ecology*, 27, 81-92.

Shaukat, S. S., Munir, N., Siddiqui, A. (2003). Allelopathic response of *Conyza canadensis* (L.) Cronquist: A Cosmopolitan Weed. *Asian Journal of Plant Science*, 2, 1034-1039.

Souto, X. C., González, L., Reigosa, M. J., (1995). Allelopathy in forest environment in Galicia, NW Spain. *Allelopathy Journal*, 2, 67-78.

Sterling, T. M., Houtz, R. L., Putnam, A. R. (1987). Phytotoxic exudates from vevetleaf (*Abutilon theophrasti*) glandular trichomes. *American Journal of Botany*, 74, 543-550.

Stjepanović-Veseličić, L., (1979). *The vegetation of Delibato Sands.* ŠIK Pančevo and the Ecological Society of Vojvodina, Novi Sad, Serbia.

Stojšić,V., Dinić, A., (2009). The dynamics and distribution of plant species in vegetation succession on the burned area of Delibato Sands. *Environmental protection* (in Serbian) 60, 305-312.

Stojšić, V., Dinić, A., Paunović, R., (2007). Monitoring of the forest ecosystems regeneration after bombing in the year 1999 in the national park Fruška Gora. Environmental protection of urban and suburban settlements: *XI international Eco-conference*, 2007, 26-29th September: proceedings. Vol. 1. 229-239. Novi Sad, Ecological Movement of the City of Novi Sad, 2007.

Tang, C. S., Young, C. C. (1982). Collection and identification of allelopathic compounds from the undisturbed root system of Bigalta Limpograss (*Hemarthria altissima*). *Plant Physiology*, 69, 155-160.

Theodoru, M. K., Gascoyne, D. J., Akin, D. E., Hartley, R. D. (1987). Effect of phenolic acids and phenolics from plant cell walls on rumenlike fermentation in consecutive batch culture. *Applied and Environmental Microbiology*, 53, 1046-1050.

Thorpe, A. S., Thelen, G. C., Diaconu, A., Callaway, R. M. (2009). Root exudate is allelopathic in invaded community but not in native community: field evidence for the novel weapons hypothesis. *Journal of Ecology*, 97, 641-645.

Timsina, B. B., Shrestha B., Rokaya M. B. and Munzbergova Z. (2011). Impact of *Parthenium hysterophorus* L., invasion on plant species composition and soil properties of grassland communities in Nepal. *Flora*, 206, 233-240.

Vinogradova, Y. K. (2007). Content of phenol compounds and silicon in plants of *Conyza canadensis* (L.) Cronquist. *Бюллетень Главного ботанического сада (Bulletin of the Main Botanical Garden)* 193, 117-127.

Vrbničanin, S., Karadžić, B., Dajić-Stevanović, Z. (2004). Adventive and invasive weed species in Serbia. *Acta biologica iugoslavica:* serija G: *Acta herbologica* 13, 1-12.

Vrbničanin, S., Stefanović L., Elezović I., Stanković-Kalezić R., Jovanović-Radovanov K., Marisavljević D., Pavlović D., Gavrić M. (2008). The distribution of some economically damaging, invasive and quarantine weed species in Serbia. Part 2: Spatial distribution and presence of nine weed species in Serbia. *Plant doctor,* XXXVI (6), 408-418.

Wagner, G. J. (1991). Secreting glandular trichomes: more than just hairs. *Plant Physiology*, 96, 675-679.

Wardle, D. A., Nilsson, M. C., Gallet, C., Zackrisson, O. (1998). An ecosystem-level perspective of allelopathy. *Biological Review*, 73, 305-319.

Weaver, S. E., (2001). The biology of Canadian weeds. 115. *Conyza canadensis. Canadian Journal of Plant Science*, 81: 867-875

Weidenhamer, J. D., Callawaym R. M. (2010). Direct and indirect effects of invasive plants on soil chemistry and ecosystem function. *Journal of Chemical Ecology*, 36, 59-69.

Westhoff, van der Marrel E. *The Braun-Blanquet approach.* In: Handbook of Vegetation Science V. Ordination and Classification of Communities. The Hague. (Ed. Whittaker, R. H.). (1973). pp. 617-726.

Whitehead, D. C., Dibb, H., Hartley, R. D., (1983). Bound phenolic compounds in water extracts of soils, plant roots and leaf litter. *Soil Biology and Biochemistry*, 15, 133-136.

Whitmore, F. (1976). Binding of ferulic acid to cell walls by peroxidases of *Pinuselliottii. Phytochemistry*, 15, 375-378.

Xingxiang, G., Mei, L., Zongjun, G., Hongjun, Z., Zuowen. S. (2009). Allelopathic effects of *Conyza canadensis* on the germination and growth of wheat, sorghum, cucumber, rape and radish. *Allelopathy Journal*, 23, 287-296.

Yang, C. M., Chang, I. F., Lin, S. J., Chou, C. H. (2004). Effects of three allelopathic phenolics on chlorophyll accumulation of rice (*Oryza sativa*) seedlings: II. Stimulation of consumption-orientation. *Botanical Bulletin of Academia Sinica*, 45, 119-125.

Yang, C. M., Lee, C. N. Chou, C. H. (2002). Effects of three allelopathic phenolics on chlorophyll accumulation of rice (*Oryza sativa*) seedlings: I. Inhibition of supply orientation. *Botanical Bulletin of Academia Sinica*, 43, 299-304.

Yoda, K., Kira, T., Ogawa, J. and Hozumi, K. (1963). Self-thinning in overcrowded pure stands under cultivated and natural conditions. *Journal of Biology* (Osaka City University), 14, 107-129.

Zaplata, M. K., Winter, S., Biemelt, D., Fischer, A., (2011). Immediate shift towards source dynamics: the pioneer species *Conyza canadensis* in an initial ecosystem. *Flora*, 206, 928-934.

In: New Developments in Allelopathy Research
Editor: Julia E. Price

ISBN: 978-1-63483-390-5

Chapter 5

ALLELOPATHIC EFFECTS OF MACROPHYTES IN SUBTROPICAL SHALLOW LAKES

Ng Haig They[1], Tiago Finkler Ferreira[2], David da Motta Marques[2], Lúcia Ribeiro Rodrigues[2], Savênia Bonoto da Silveira[3], Alessandra de Abreu Arriada[3], Luciane Oliveira Crossetti[2], Luciana de Souza Cardoso[2] and Carlos Ruberto Fragoso Jr.[4]

[1]Universidade Federal do Rio Grande do Norte - UFRN, Brazil
[2]Universidade Federal do Rio Grande do Sul – UFRGS, Brazil
[3]Universidade Federal do Rio Grande – FURG, Brazil
[4]Universidade Federal de Alagoas – UFAL, Brazil

ABSTRACT

In subtropical shallow lakes, due to shallowness and mild climate, macrophytes grow continuously and colonize extensive areas. Many macrophytes are known for the production and release of allelochemicals, with effects extensively reported for phytoplankton and cyanobacteria, but poorly elucidated for heterotrophic bacteria. In this chapter, we reunited three studies conducted in subtropical shallow lakes in southern Brazil, where many allelopathic macrophytes occur, such as *Potamogeton illinoensis, Potamogeton pectinatus, Cabomba caroliniana, Egeria densa, Myriophyllum spicatum* and *Ceratophyllum demersum*. In the first study, the phytoplankton community and environmental variables were assessed in 5 seasonal campaigns through transects ranging from a macrophyte-vegetated to a macrophyte-free area in the large Lake Mangueira. The chlorophyll *a*, cyanobacterial biomass, the total biomass, diversity and richness of the phytoplankton community, as well as several nutrients were lower in the macrophytes dominated part of the transect and increased in the macrophytes-free area. In the second study, microcosm coexistence experiments with *C. caroliniana, E. densa, M. spicatum,* and *C. demersum*) and the cyanobacteria *Microcystis aeruginosa* under different levels of phosphorus concentrations have revealed that, although all plants were effective competitors for nutrients and impacted negatively the *M. aeruginosa* biomass, only *M. spicatum* and *C. caroliniana* provided evidence of allelopathic inhibition. In the third study, the allelopathic effects of *P. pectinatus* on the microbial food web (heterotrophic bacteria, flagellates and ciliates) and phosphatase activity were measured in the field in areas with and without the macrophytes and also in mesocosm experiments with the addition of methanolic extracts, decomposing plant parts and water surrounding the stands. Field results showed lower microbial food web components densities as well as enzymatic activity in areas with the presence of the macrophytes, whereas in the experiments the methanolic extract and the decomposing stems and leaves induced a one day longer lag phase in the bacterial growth and lower rates of enzymatic activity when compared with the control. Taken together, these results provide further field evidence for the allelopathic effects of macrophytes on phytoplankton, autotrophic and heterotrophic bacteria, corroborating much of the literature available so far in this topic, as well as provide a mechanistic explanation for the lower bacterial density, biovolume, biomass, production and respiration often found in areas where these macrophytes occur. Allelopathy adds, hence, another layer of complexity to the already known strong structuring role of macrophytes in shallow lakes, imparting poorer functioning of microbial food webs in areas densely colonized by these plants.

INTRODUCTION

Even though the phenomenon of allelopathy has been recognized over two thousand years, the concept was coined by Hans Molisch only in 1937 (Molisch, 1937) from the greek words *allelon* (= mutual, among each other) and *pathos* (= suffering or feeling) (Willis, 2007). Originally intended to describe simply the effect of one plant on another, either positive or negative, the concept of allelopathy has evolved through time with much controversy and mostly linked with negative effects (Willis, 2007). Over the time the concept has broaden and narrowed, to the point that zoologists have relatively recently borrowed the term to address chemical interactions among animals (Gross, 1999) (see a complete review on this topic in Willis, 2007). Modern concepts also include the release of chemicals that potentially affect other plants from dead plant material, (including dead litter) (Wardle et al., 1998). As we will see along this chapter, considering dead plant material is crucial for understanding allelopathic effects on microorganisms in the subtropical shallow lakes considered.

The history of the field of allelopathy reveals that most of it has focused on terrestrial ecosystems, particularly on the interaction between cultivated species (Willis, 2007). In aquatic ecosystems, the allelochemicals can be produced by eukaryotic algae, cyanobacteria or angiosperms (Gross, 2003) and, contrary to terrestrial ecosystems, can potentially affect much further organisms and many trophic levels because of higher capacity of dispersion in the water.

Allelopathy in aquatic ecosystems has always been intriguing as a phenomenon explaining the mutual dominance of either submerged macrophytes or phytoplankton in shallow lakes (Gross et al., 2007). Although phytoplankton and epiphyte densities in these systems is the result of complex abiotic and biotic feedback mechanisms, allelopathy is considered as one alternative mechanism (Scheffer et al., 1993) at high (i.e., non-limiting) nutrient levels, at least when known allelopathic plant species and/or sensitive phytoplankton species are involved (Korner & Nicklisch, 2002; van Donk & van de Bund, 2002). Thus, macrophytes are the major players when it comes to the production of allelopathic compounds in shallow lakes.

Allelochemicals excreted by aquatic macrophytes could be an effective defense strategy against other photosynthetic organisms competing for light and nutrients, like epiphyton and phytoplankton (Gopal and Goel, 1993; Elakovich and Wooten, 1995). The identified allelochemicals belong to rather different chemical classes such as sulphur compounds, polyacetylenes, polyphenols and oxygenated fatty acids (Gross, 1999). In this scenario, the degradation of allelochemicals by heterotrophic bacteria adds another layer of complexity governing the outcome of the allelopathic interactions among algae species and between algae and macrophytes (Hulot and Huisman, 2004). However, direct allelopathic effects on heterotrophic bacteria remain unclear and unacknowledged, and one of this chapter aims is casting some light into this topic.

Proving evidence for allelopathy, however, remains difficult. According to Willis (1985), six pre-requisites have to be met to unequivocally show the occurrence of allelopathy: i) a pattern of inhibition of target plant(s) or alga(e); i) detection of allelopathic compound(s) produced by donor plants; iii) release of these compounds by the producing plant; iv) transport and/or accumulation of the allelopathic compounds in the environment; v) uptake by the target organism(s); and vi) the inhibition cannot be explained solely by other physical or biotic factors, especially herbivory and competition.

Even though several reports of macrophytes that can produce allelopathic substances affecting phytoplankton and periphyton (Jasser, 1995; van Donk and van de Bund, 2002; Kluijver et al., 2015), and higher trophic levels (Lauridsen and Lodge, 1996; Burks et al., 2000; Della-Greca et al., 2003) exist, the ecological interactions that macrophytes are involved with can masquerade the possible allelopathic effects.

Macrophytes are of central role in structuring shallow lakes' food webs (Jeppesen et al., 1998; Scheffer, 1998; Meerhoff et al., 2007; Rodrigues et al., 2013). Examples include: i) competition for nutrients and other resources with phytoplankton and periphyton (van Donk et al., 1993; van Donk and van de Bund, 2002; Kluijver et al., 2015); ii) reduced mixing of the water column (which increases phytoplankton sinking losses) and reduced light climate for phytoplankton (Barko and James, 1998; Horppila and Nurminen, 2003; 2005); iii) increased denitrification inside macrophyte beds (e.g., Weisner et al., 1994); iv) provision of structure and shelter for organisms, with very specific communities of periphyton and plant-associated invertebrates inhabiting the plants' surfaces (Lauridsen et al., 1996; Rodrigues et al., 2013); v) provision of shelter against predators for zooplankton (e.g., Schriver et al., 1995; Meerhoff et al., 2007; Gazulha et al., 2011) and small fish (Persson and Eklov, 1995; Garcia et al., 2006; Teixeira de Melo et al., 2009); and vi) being one of the main factors structuring fish communities in shallow lakes (Lammens, 1989; Meerhoff et al., 2007; Teixeira de Melo et al., 2009), largely due to differences between fish species in exploring food sources (Persson et al., 1993) and differences in the use of macrophytes as a substrate for egg deposition.

The activity of allelopathic compounds by macrophytes adds a new dimension to the already well documented ecological effects of these plants summarized above. The literature concerning isolation and testing allelopathic effects of macrophytes on several organisms in mesocosms and field surveys is abundant. For example, a sulphur compound with allelopathic properties has been isolated from *Ceratophyllum demersum* Wium-Andersen et al. (1983). Later on, Mjelde and Faafeng (1997) found that *C. demersum* hampers phytoplankton development in small Norwegian lakes over a wide range of phosphorus concentrations and geographical latitude. Also van Donk and Gulati (1995) suggested that *C. demersum* can inhibit the growth of both epiphytes and phytoplankton.

Wium-Andersen et al. (1982) also isolated two potentially allelopathic compounds, a dithiolane and a trithiane, from both fresh and brackish water characean species. These substances and elemental sulphur, often found when trithiane is extracted from the plants, inhibited the growth of the diatom *Nitzschia palea*. Whether *Chara* spp. inhibit phytoplankton growth in the field due to these allelopathic substances is still not clear, as the presence of these substances within plants is no proof of their release and activity in the environment (Wium-Andersen, 1987). Blindow and Hootsmans (1991) also found a differential sensitivity among phytoplankton species for allelopathic substances. The growth of the green alga *Ankistrodesmus* was stimulated or not affected by water originated from *Chara* cultures, while the growth of the green alga *Scenedesmus* was significantly reduced by cultures of *Chara globularis* with about 10% addition. The excretion of allelopathic substances by some of these macrophyte cultures changed with time.

Allelopathic effects of *Chara* on zooplankton also are not well known, although some studies do report such effects by other macrophytes. Dorgelo and Heykoop (1985) found an avoidance of *Potamogeton lucens* and *Elodea nuttallii* by *Daphnia longispina* in the light and they suggested that the antagonistic effect of the plants is less active in the absence of active photosynthesis; no effect was found with artificial plants. Burks et al. (2000) suggested that prolonged exposure of *Daphnia* to exudates from the macrophyte *Elodea canadensis* incurs costs for these animals. Daphnids exposed to *Elodea* exudates took longer to mature and possessed fewer eggs than daphnids not exposed to *Elodea*.

Lauridsen and Lodge (1996) confirmed the repellent effect of plants but showed that it can be overriden by fish-related stimuli. In aquaria that were partly planted with *Myriophyllum*, a majority of the animals stayed in the open water. However, when fish was introduced most of the animals preferred the vegetated parts. Plastic plants had the same effect, although not as strong as the real ones, indicating that not only plant odor but also the structure itself repels the animals. On the other hand, adding fish odor works as strongly as adding caged fish, suggesting that it is largely the chemical cue that drives *Daphnia* into the refuges. Burns and Dodds (1999) stated that the possible effects of allelopathy from macrophytes on zooplankton calls for further studies in ecosystems. However, the existence of very abundant specialized zooplankton communities in macrophyte beds indicates that negative effects of macrophytes on zooplankton is not likely to be very important, but this is far from being a completely elucidated topic.

Some authors use plant homogenates or extracts to test for allelopathy (Gross et al., 2007). Extract-based methods offer the chance to compare the allelopathic potential of different macrophyte species or seasonal differences in one species in a standardised way. The direct application of extracts also enables comparison of the response of nutrient limited and non-limited phytoplankton in order to test whether additional stress of target organisms results in a higher sensitivity against allelopathic substances (Reigosa et al., 1999). Additionally, the producing plants can be cultivated under different nutrient and light conditions to study the impact of resource availability on the production of allelopathic compounds (Gross, 2003).

Bioassay-directed extractions of *Ceratophyllum demersum*, *Elodea* spp., *Myriophyllum* spp., *Najas marina* and *Stratiotes aloides* with different solvents and solvent mixtures (water, methanol, and acetone; pure or mixed with water) often reveal the highest activity against cyanobacteria when solvents with intermediate lipophilic properties were used, e.g., 50% acetone or 70% methanol in water (Gross et al., 1996, 2003; Erhard & Gross, 2006; Mulderij et al., 2007). Under these conditions both hydrophilic and moderately lipophilic compounds can be extracted from the plant tissue. This method will result in compounds that in general have a low molecular weight, are fairly water-soluble, and can bind to and/or pass through cell membranes.

Experiments conducted with extracts certainly give no convincing ecological proof for allelopathic interactions. Such studies should preferably be presented accompanied with experiments using culture filtrate (exudate) from the active macrophytes (Gross et al., 2007). One of the factors supporting the use of algae and invertebrates in the bioassays is their trophic level position in aquatic ecosystems. If organisms at the bottom of the food chains, e.g., algae and some invertebrates, suffer disturbances from toxicants such as some allelochemicals, the ecosystemic effects may be stronger.

Future research in analyzing the presence of allelopathic substances should focus on studies with intact macrophytes and experimental setups in which it is possible to make a clear distinction between allelopathy and other growth limiting factors. Furthermore, it should analyze the role of allelopathy under natural conditions and to determine the importance of this factor in the complex ecosystem context.

In this chapter we elucidate many of these issues raised in the preceding paragraphs concerning the ecological repercussions of the allelopathic effects of macrophytes in subtropical shallow lakes. In these systems, given the mild climate conditions and shallowness, macrophytes are able to grow continuously throughout the year and colonize extensive areas within the lakes, imposing permanent and strong ecological repercussions. The ecosystems approached in this chapter are colonized by several macrophytes with reported allelopathic effects. In the first section (Phytoplankton and Autotrophic Bacteria) we demonstrate strong phytoplankton community changes in composition and biomass in response to the presence of macrophytes in the field and also the effects of four macrophytes on the growth of a cyanobacteria under different nutrient levels through coexistence experiments. In the second section (Heterotrophic Bacteria), we present experimental evidence for a direct allelopathic effect of macrophytes on heterotrophic bacteria through reduced enzymatic activity, which provides a mechanistic explanation of the lower density, biomass, biovolume, secondary production and respiration of heterotrophic bacteria found associated with macrophytes's stands in these lakes.

Although not strictly fulfilling all the requirements enumerated by Willis (1985) for the unequivocal demonstration of allelopathy, we intend to advance on this rather overlooked topic which we believe has strong implications for the ecology of not only subtropical shallow lakes, but all lakes where macrophytes with allelopathic properties grow.

Phytoplankton and Autotrophic Bacteria

It is well known that submerged macrophytes increase water transparency through positive physical feedback mechanisms and negative chemical interference with phytoplankton (Scheffer, 1998; Gross et al., 2007; Lürling et al., 2006). Several mechanisms are involved in the antagonistic relationship between phytoplankton and macrophytes (Scheffer et al., 2001; Van Nes et al., 2002, 2003). Among these, allelopathy has been evoked as one of the steering mechanisms behind the mutual dominance of either submerged macrophytes or phytoplankton in shallow lakes (Hootsmans and Blindow, 1994; Søndergaard and Moss, 1988; Gross et al., 2007).

Since the allelopathy of macrophytes has been confirmed experimentally, it has become a potential alternative to suppress the growth of undesirable algae, especially cyanobacteria, in eutrophic ecosystems (Nakai et al., 2000). In addition, competition for nutrients has also been addressed as an effective indirect way of controlling algal growth (van Donk et al., 1993; Kufel and Kufel, 2002). Both allelopathy and nutrient competition are considered as positive feedback mechanisms exerted by macrophytes to inhibit or reduce phytoplankton growth, contributing to the maintenance of low phytoplankton biomass and stabilizing the clear oligotrophic water state within macrophyte beds (Scheffer, 1998; Van den Berg et al., 1998).

The release of allelopathically active compounds such as the algicidal polyphenol tellimagrandin II by *Myriophyllum spicatum* has been confirmed previously (Gross et al., 1996; Nakai et al., 2000; Leu et al., 2002) as having a strong inhibitory effect against various cyanobacteria and to a lesser extent on chlorophytes and diatoms (Gross et al., 1996; Körner and Nicklish, 2002). Other species of macrophytes such as *Chara globularis* and *Elodea canadensis* have been shown to reduce the growth of the chlorophyte *Scenedesmus obliquus* (Lürling et al., 2006). The allelopathic activity of *Ceratophyllum demersum* (Körner and Nicklish, 2002; Nakai et al., 2000) and *Stratiotes aloides* (Mulderij et al., 2007) was also verified against cyanobacteria.

Although several species of submerged macrophytes may contain allelochemicals that suppress algal growth, the interference of nutrients and light is one of the main complications in investigations of allelopathic interactions (Hilt et al., 2006; Gross et al., 2007). Nutrient availability may impact resource allocation of macrophytes for growth versus production of allelochemicals, as well as the competition between macrophytes and phytoplankton (Gross, 2003b). In summary, environmental variables such as light, temperature, and eutrophication levels may influence the interactions among species under particular ecosystem conditions (Gross et al. 2007).

The fact that submerged plants are able to prevent excessive phytoplankton growth (Blindow et al., 1993; Van den Berg et al., 1998) causing the clear state to be self-stabilizing is a well established theory (Moss, 1990; Scheffer et al., 2001; van Nes et al., 2003). However, the confirmation of allelopathy as one of the mechanistic explanations is still speculative at the ecosystem level, i.e., in the field (Hilt et al., 2006; Lurling et al., 2006) even considering that allelochemicals have been detected, extracted and tested in laboratory in the last years. In the field, most of the findings have been carried out in temperate lakes, where there is a seasonal die-off of the plants, differently from continuous growth in subtropical and tropical regions. In such systems, once a submerged vegetation stand is established, it can be expected to display continuous growth and production of allelochemicals throughout the year. Thus, allelopathy can potentially enhance resilience of the clear water state. These are very interesting gaps which have important implications for management and restoration that are now being addressed by researchers in subtropical lakes in Southern Brazil.

In order to understand the allelochemical potential of macrophytes in changing phytoplankton community composition and reducing cyanobacterial growth in subtropical regions, we have carried out field and experimental approaches (Finkler Ferreira, 2009). Initially we present data from field monitoring in a large subtropical lake Mangueira, where differences in the phytoplankton community composition, especially cyanobacteria, were clearly observed along a littoral macrophytedominated to a pelagic phytoplankton-dominated gradient. Such findings led us to test the hypothesis of allelopathy and nutrient competition experimentally in laboratory. These approaches are summarized here separately and their conclusions are integrated at the end of this section. The complete information from these works is available in Finkler Ferreira (2009).

Field Assessment: The Structuring Role of Macrophytes on the Phytoplankton Community in a Large Subtropical Lake

Few studies have focused on phytoplankton in relation to the presence of submerged macrophytes along a littoral-pelagic gradient. The aim of this study was to evaluate the effects of a densely submerged macrophyte bed on water quality and phytoplankton community structure along a littoral (macrophyte dominated) to pelagic gradient (phytoplankton dominated) in a large shallow lake-wetland ecosystem, in southern Brazil. In this work, Finkler Ferreira (2009) showed the role of subtropical submerged macrophytes on the nutrient pool and structuring effects on phytoplankton, indicating traces of allelopathy and nutrient competition with cyanobacteria.

Material and Methods

Study Site

The study was carried out in the large shallow lake Mangueira (max. depth 6 m). This lake is situated in southern Brazil along the Atlantic Ocean (33°31'22"S 53°07'48"W) and is 90 km long and 3-10 km wide (Figure 1). This ecosystem was originated from the last glaciation event on the southeast coast of South America (during the Pleistocene ~5,000 years ago). Despite its proximity to the Atlantic Ocean (≅ 3 km), salt sprays do not affect the salinity of the lake.

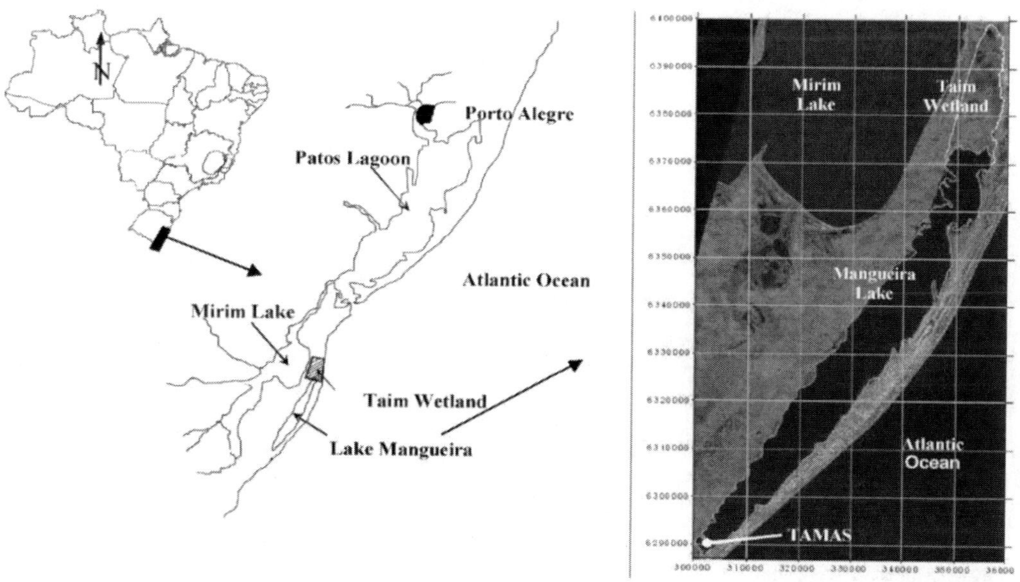

Figure 1. Location of Lake Mangueira, southern Brazil. The study was conducted in the southern part of the lake, nearby a weather station named TAMAS.

Along the lake's shoreline, there are large areas colonized by dense belts of emergent macrophytes that reduce water currents, allowing the establishment of submerged macrophyte communities, which may reach up to 6.0 m depth.

The lake is hydrologically linked to two large wetlands, one in the northernmost and other in the southernmost part. Together, they constitute a synergetic system driven by intense winds (SW-NE). The production and decomposition of the macrophytes in the wetlands provide large amounts of humic substances that color the water nearby the wetlands.

Sampling along the Macrophyte-Pelagic Gradient

In order to assess the effect of the submerged macrophytes on the water quality and phytoplankton community structure, samplings were undertaken along a transect ranging from a densely submerged vegetated littoral to an unvegetated pelagic zone, in the southernmost part of the lake Mangueira. In the vegetated area, the margin was covered by emergent macrophytes being surrounded by a large wetland belt. The transect was approximately 3,220 m, comprising eight sampling points. The distribution of the points was not regular, but arbitrarily chosen to cover the heterogeneity of the habitats along the macrophyte-pelagic gradient (Figure 2).

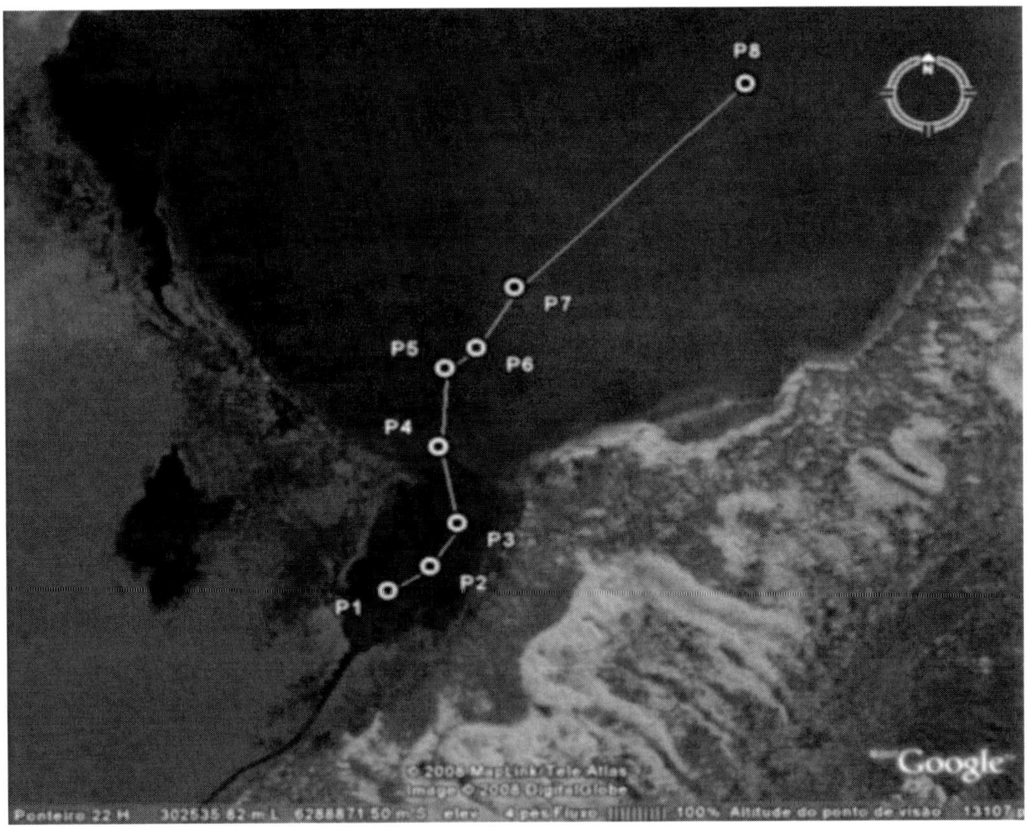

Figure 2. Map of the southern part of Lake Mangueira. In evidence: the sampling points along the transect (~3,220 m) established from a submerged vegetated littoral to an unvegetated pelagic zone.

Samplings were carried out seasonally. Five campaigns were conducted from May 2007 until March 2008 it total. During the samplings, analysis of the submerged macrophytes community and immediate measurements of limnological variables were carried out in the field.

Concomitantly, water samples were collected for nutrients, phytoplankton and zooplankton analysis in the laboratory. The abundance of submerged macrophytes was estimated as the percentage of volume infestation (PVI) under water (Canfield et al., 1984). The calculus of PVI was done through the following equation:

$$PVI\ (\%) = \frac{MacC \times MacL}{D} \tag{1}$$

Where, MacC is the macrophyte coverage (%) of each species, MacL is the macrophyte length (m) and D is the depth (m). The macrophyte coverage of each species was estimated visually as the percentage covered by each species within plots of 9.0 m^2.

In the field, yellow substances were estimated using a submergible espectrofluorometer (BBE-Moldaenke). Phytoplankton was quantitatively estimated according to Utermöhl (1958) with sedimentation time following Lund et al. (1958). Biomass (mg L^{-1}) was estimated by multiplying the biovolume by density of each taxon and applying empirical conversion factors from biovolume to biomass (Wetzel and Likens, 2000), considering, whenever possible, the mean dimension of 20 individuals. Diversity index followed Shannon and Weaver (1963) and evenness by Lloyd and Ghelardi (1964). Phytoplanktonic species were assigned into functional groups according to Reynolds et al. (2002) and Padisák et al. (2009). C-R-S strategies classification followed Reynolds (1997).

Results

Along the transect, PVI decreased linearly with increasing depth (Table 1). Inside the macrophyte bed (sampling points P1 and P2), the plants reached the water surface during all seasons, with PVI values varying between 94-100%. From the sampling point P3 and further out, the vegetation became sparse and PVI declined over a short distance until reaching values around 5-15% at P5. Therefore the sampling point P5 was considered the interface zone between the macrophyte dominated and pelagic dominated area.

The dense vegetation bed was composed by seven dominant species: *Potamogeton illinoensis*, *Cabomba caroliniana*, *Egeria densa*, *Myriophyllum spicatum*, *Nitella* sp., *Potamogeton pectinatus*, *Ceratophyllum demersum* and *Utricularia* sp. Within the bed (P1-P3), the species with higher mean PVI over the sampling campaigns were: *E. densa* (25.6%), *M. spicatum* (22.3%), *C. caroliniana* (21.6%), and *P. illinoensis* (19.8%). The environmental structure changed along the transect from stagnated waters, protected by emergent macrophytes, to a more turbulent interface zone. The general succession of species from P1 to P5 was observed in the following sequence: *P. pectinatus* - *Utricularia* sp., *C. caroliniana*, *M. spicatum*, *E. densa*, *P. illinoensis*. At the interface zone, *P. illinoensis* was the only species observed colonizing the deeper areas (Table 1).

Table 1. Percentage of Volume Infested (PVI) and schematic illustration of the submerged macrophyte species along the transect in Lake Mangueira

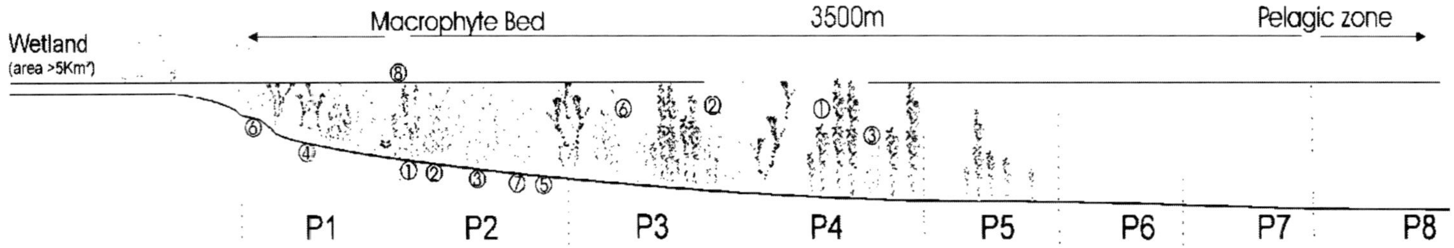

Season	Species	P1	P2	P3	P4	P5	P6	P7	P8
Autumn (May/2007)	Potamogeton illinoensis	11.6	25.0	36.67	19.23	5.71	-	-	-
	Cabomba caroliniana	20.8	21.7	-	-	-	-	-	-
	Egeria densa	20.8	25.3	-	-	-	-	-	-
	Miriophyllum	23.6	10.0	-	-	-	-	-	-
	Nitella spp.	-	2.2	-	-	-	-	-	-
	Potamogeton pectinatus	10.0	3.5	-	-	-	-	-	-
	Utricularia spp.	-	2.3	-	-	-	-	-	-
	Total	86.8	100	36.67	19.23	5.71	-	-	-
Winter (July/2007)	Potamogeton illinoensis	10.0	2.6	20.7	18.3	16.1	-	-	-
	Cabomba caroliniana	32.5	9.3	-	-	-	-	-	-
	Egeria densa	8.6	62.7	-	-	-	-	-	-
	Miriophyllum	35.0	9.3	-	-	-	-	-	-
	Nitella spp.	-	0.5	-	-	-	-	-	-
	Potamogeton pectinatus	10.0	-	-	-	-	-	-	-
	Ceratophyllum demersum	-	13.9	-	-	-	-	-	-
	Total	96.1	98.3	20.7	18.3	16.1	-	-	-

Spring (Sep/2007)	Potamogeton illinoensis	10.00	10.67	6.88	3.14	-	-	-	-
	Cabomba caroliniana	32.50	16.67	-	-	-	-	-	-
	Egeria densa	8.57	19.33	-	-	-	-	-	-
	Miriophyllum	35.00	46.67	-	-	-	-	-	-
	Nitella spp.	-	2.00	-	-	-	-	-	-
	Potamogeton pectinatus	10.00	-	-	-	-	-	-	-
	Utricularia spp.	-	2.30	-	-	-	-	-	-
	Total	96.07	99.63	6.88	3.14	-	-	-	-

Summer (Dec/2007)	Potamogeton illinoensis	9.4	-	20.8	2.4	13.3	-	-	-
	Cabomba caroliniana	20.8	24.1	-	-	-	-	-	-
	Egeria densa	19.4	50.0	-	4.8	-	-	-	-
	Miriophyllum	23.6	14.8	-	-	-	-	-	-
	Nitella spp.	-	5.2	-	-	-	-	-	-
	Potamogeton pectinatus	25.0	-	-	-	-	-	-	-
	Ceratophyllum demersum	-	5.9	-	-	-	-	-	-
	Total	98.3	100.0	20.8	7.2	13.3	-	-	-

Sum-Aut (Mar-Apr/2008)	Potamogeton illinoensis	35.0	25.4	43.2	4.6	8.6	-	-	-
	Cabomba caroliniana	15.0	22.9	-	-	-	-	-	-
	Egeria densa	32.7	24.7	6.4	9.2	2.1	-	-	-
	Miriophyllum	15	19.4	3.2	2.7	-	-	-	-
	Nitella spp.	-	-	6.8	-	-	-	-	-
	Potamogeton pectinatus	1.6	-	-	-	-	-	-	-
	Utricularia spp.	-	2.0	-	-	-	-	-	-
	Total	99.3	94.4	59.5	16.5	10.7	-	-	-

Legend of the macrophyte species at the illustration: (1) *Potamogeton illinoensis*, (2) *Cabomba caroliniana*, (3) *Egeria densa*, (4) *Myriophyllum spicatum*, (5) *Nitella* spp., (6) *Potamogeton pectinatus*, (7) *Ceratophyllum demersum*, (8) *Utricularia* spp.

Water Quality Gradient along the Transect

Over all sampling seasons, clear differences in water quality variables were observed along the submerged macrophyte-pelagic gradient. Among the variables assessed, total phosphorous (TP), ortophosphate (PO_4^{-3}-P), bicarbonate (HCO^{-3}), inorganic carbon (IC), silicate and chlorophyll *a* (chla) showed a negative correlation with PVI, increasing outwards the vegetation bed. Along the transect, the concentration of chla varied significantly ($F(7,16) = 6.999$; $p < 0.001$). The values remained relatively low at all seasons, ranging from 0.74 to 4.6 μg L^{-1}. The peak of chlorophyll *a* was observed in spring (September/07). This variable also presented a significant negative correlation with yellow substances ($R^2 = -0.41$; $p < 0.01$). With respect to nutrients, the concentration of PO_4^{-3}-P was very low during all seasons, varying between 0.010 to 0.038 mg L^{-1}. Although the mean concentration of PO_4^{-3} was relatively low, its variation was statistically significant ($F(7,16) = 3.689$; $p < 0.01$) along the transect, considering all campaigns. Nitrate (NO_3^--N) and ammonium (NH_4^+-N) were observed varying at low concentrations (0.014-0.037 and 0.004-0.017 mg L^{-1}, respectively) and did not show any particular trend along the transect. On the other hand, bicarbonate (HCO_3^-) showed a clear and significant increase towards the pelagic zone, from 33.5 to 77.0 mg L^{-1} ($F(7,16) = 8.344$; $p < 0.001$). The variables related to the presence of organic matter such as dissolved organic carbon (DOC) and yellow substances decreased linearly with the PVI towards the pelagic area ($R^2 = 0.922$; $P < 0.001$) at all sampling dates.

Vertical light attenuation coefficient (K_d) was relatively low (1-3 m^{-1}) along the transect at all samplings campaigns, being predicted by the concentration of yellow substances, dissolved organic carbon (DOC) and chlorophyll *a* (Equations 2, 3, 4). However, these variables were better predictors of light extinction, independently, at distinct zones of the transect. Within the macrophyte bed, Kd was significantly related to the concentration of yellow substances and dissolved organic carbon (DOC) (Equation 3), whereas in the pelagic zone (P6 - P8) chla was the main predictor (Equation 4).

$$K_d = -0.946 + (0.323 \times YS) + (0.0942 \times DOC) + (0.472 \times Chl - a)$$

$N = 39$ Adj.$R^2 = 0.32$ $P > 0.001$ (2)

$$K_d = -1.451 + (0.846 \times YS) + (0.0578 \times DOC)$$

$N = 20$ Adj-$R^2 = 0.43$ $P < 0.001$ (3)

$$K_d = 0.0380 + (-0.490 \times Chl - a)$$

$N = 14$ Adj.$R^2 = 0.37$ $P < 0.05$ (4)

Where YS stands for yellow substances, DOC for dissolved organic carbon and chla for chlorophyll *a*. The mean values of several limnological variables differed significantly between the two distinct areas of the transect: sampling points P1 to P4 (macrophyte bed) and P6 to P8 (pelagic area). The ions PO_4^{-3}-P and HCO_3^- were significantly higher in the pelagic area. On the other hand, the concentration of DOC was significantly lower, being followed by yellow substances. Both TP and Total Solids (TS) were significantly higher in the pelagic area, whereas the pool of TN and its forms NO_3^--N and NH_4^+-N did not show any clear difference.

Phytoplankton Structure

Total phytoplanktonic biomass was considered significantly different ($F(7,16) = 8.06$; $p < 0.01$) between groups of sampling points in the macrophyte-littoral and pelagic areas. Biomass values increased along the transect, being higher in the pelagic zone (P6 - P8) and reaching 15.9 mg L^{-1} at P7 in June/07. Lower values of biomass were registered in the macrophyte-littoral region, where the average value was 1.45 mg L^{-1} (N= 18) (Figure 3).

Specific richness' and diversity varied among sampling points and a slight increase was observed towards the pelagic zone direction. Highest specific richness was found at P8 in April/08 (35 taxa) and the lowest ones were registered at P1 (July/07) and P2 (April/08), with 10 and 13 species, respectively. Higher diversity was found at P5 (September/07 April/08) and P8 (April/08), with values of 4.3, 3.9 and 3.9 bits mg^{-1}, respectively (Figure 3).

Bacillariophyceae was the most representative algae class within the macrophyte beds, contributing with 41% of mean biomass, whereas the Cyanobacteria class contributed more (43%) in the pelagic region (P6 – P8). Cyanobacteria was responsible for the increase of biomass values along the transect, presenting strong positive and significant correlations with total biomass concentration ($R^2 = 0.80$; $p < 0.001$). A significant five-fold increase of cyanobacteria mean biomass along the transect was registered ($F(7,16) = 16.4$; $p < 0.001$) (Figure 4). In Figure 4 it is shown the distribution of cyanobacteria and other algae groups throughout the macrophyte-pelagic transect at each sampling date.

The pool of total phosphorous (TP) showed a clear positive correlation with chla ($R^2 = 0.905$; $P < 0.001$) throughout the macrophyte-pelagic gradient. The mean phytoplankton biomass was negatively correlated with humic substances ($R^2 = -0.786$; $P = 0.014$). In contrast, the increase of phytoplankton biomass towards the pelagic region was related to the increase in TP and HCO_3^- (Figure 5A). The cyanobacteria biomass was only higher where PVI and concentration of YS were lower (Figure 5B).

$$\boldsymbol{Phyto = -1.906 + (212.499 \times TP) + (0.00386 \times HCO_3^-)}$$

N = 39, Adj. $R^2 = 0.164$ ($P < 0.05$)

$$\boldsymbol{Cyano = -2.164 - (0.542 \times YS) - (0.00449 \times PVI)}$$

N = 39, Adj. $R^2 = 0.287$ ($P < 0.01$)

This was corroborated by the highly negative correlation of cyanobacteria with PVI and YS values ($R^2 = -0.970$ and $R^2 = -0.881$; $P < 0.001$). On the other hand, there was a positive relationship between cyanobacteria and the concentration of nutrients as $PO_4^{-3}P$ and HCO_3^- ($R^2 = 0.881$ and $R^2 = 0.976$, $P < 0.001$). These variables showed to predict well the cyanobacteria biomass (Equation 5)

$$\boldsymbol{log(Cyano) = 0.890 - (1.310 \times log(YS)) - (0.00682 \times PVI) - \left(0.700 \times log\left(PO_4^{-3}\right)\right) - (1.251 \times log(HCO_3^-))} \quad (5)$$

N = 39 Adj. $R^2 = 0.628$ $P<0.001$

Where, Cyano is cyanobacteria biomass (mg L^{-1}), YS is yellow substances (r.u), PO_4^{-3} is orthophosphate (mg L^{-1}) and HCO_3^- is bicarbonate (mg L^{-1})

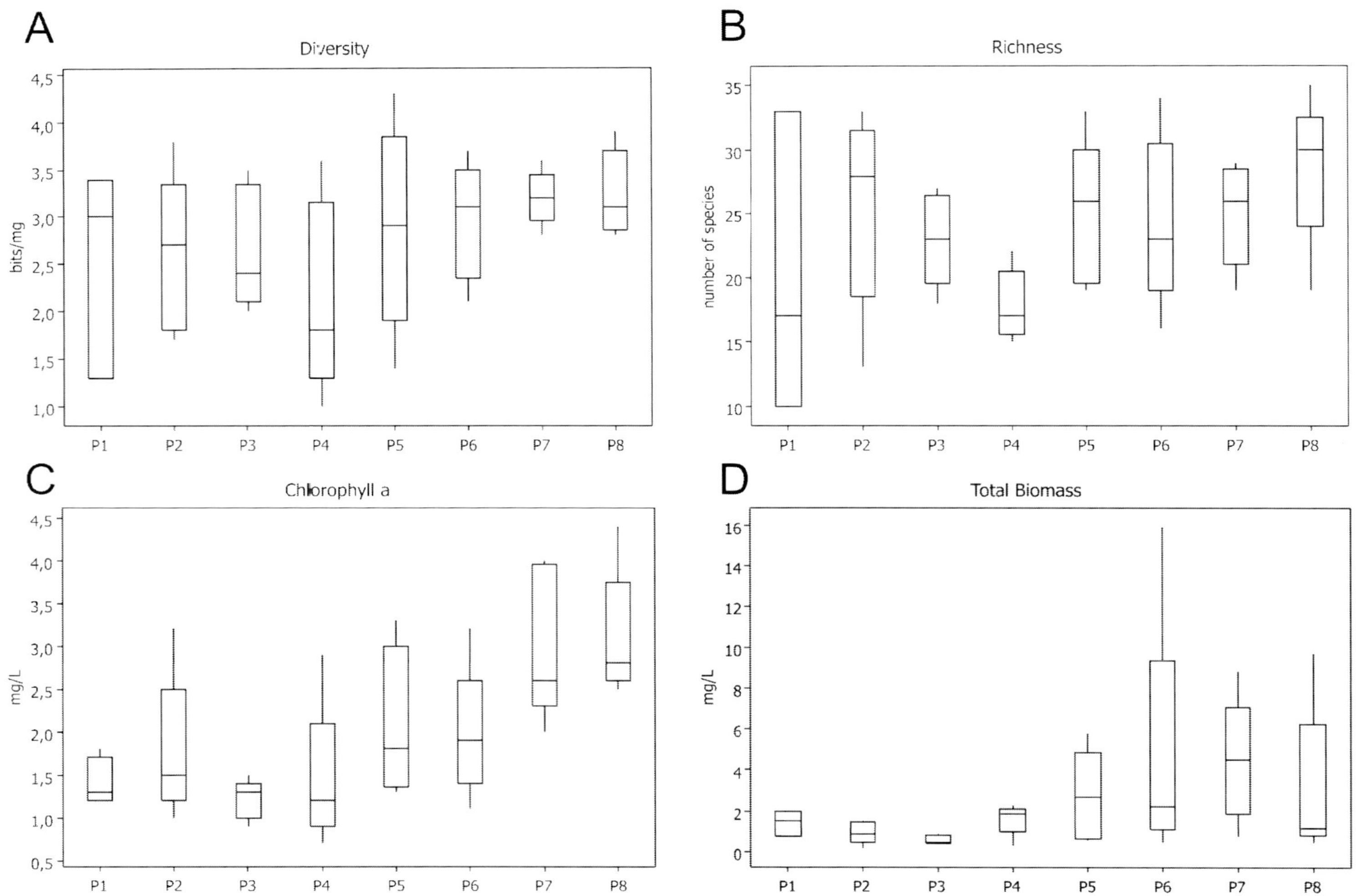

Figure 3. Boxplot (median, interquartile ranges, n = 3 for P1, n = 5 for P2-P8) of (A) Shannon – Weaver diversity (bits mg^{-1}), (B) richness (number of species), (C) chlorophyll *a* (mg L^{-1}) and (D) total biomass of phytoplankton (mg L^{-1}) along the transect in Lake Mangueira.

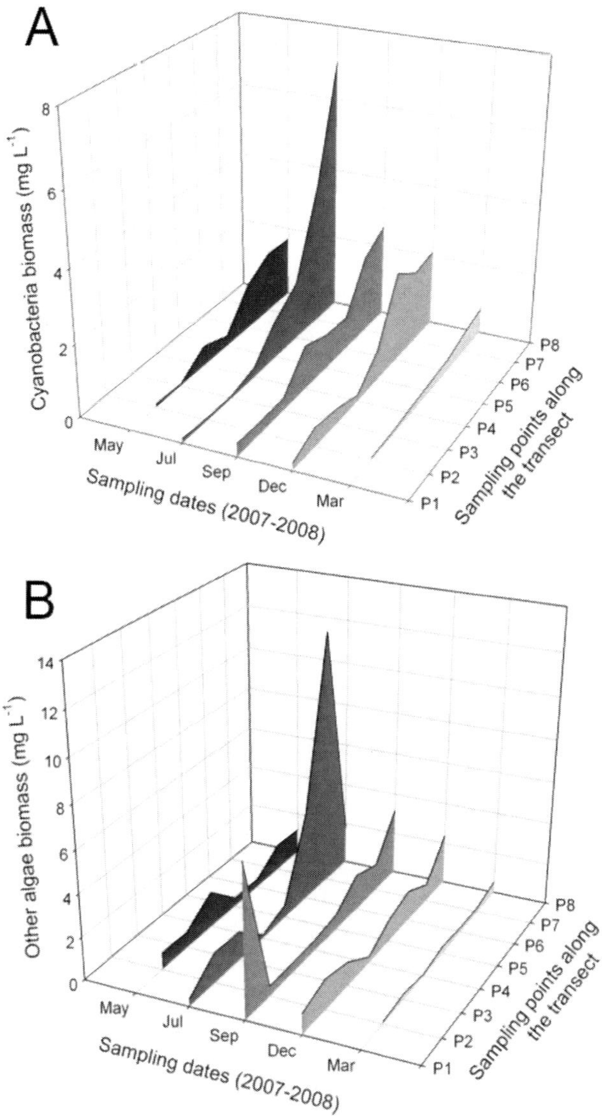

Figure 4. Biomass (mg L^{-1}) of cyanobacteria (A) and other algae groups (B) along the transect at all sampling dates in Lake Mangueira, Brazil.

Discussion

The submerged macrophytes clearly influenced the pool of nutrients and other limnological variables within and beyond the extent of the vegetation bed. In one hand, nutrients as orthophosphate (PO_4^-), bicarbonate (HCO_3^-) and the total phosphorous (TP) increased significantly outwards the macrophyte bed. On the other hand, yellow substances (YS) and dissolved organic carbon decreased in the pelagic zone. This is explained by the fact that macrophytes affect the nutrient cycling in shallow lakes, acting as nutrient sinks in several ways (Carpenter, 1981). Significant changes in the phytoplankton community structure were found along the macrophyte-pelagic gradient. Many are the aspects related to the influence of submerged beds on phytoplankton structure, as macrophytes are i) competitors for many resources, such as nutrients and light; ii) refuge for plant-associated organisms and small fishes, which may predate phytoplankton; iii) promoters of water stability, reducing mixing and resuspension and increasing sinking; and iv) producers of allelopathic substances, which may poison phytoplankton.

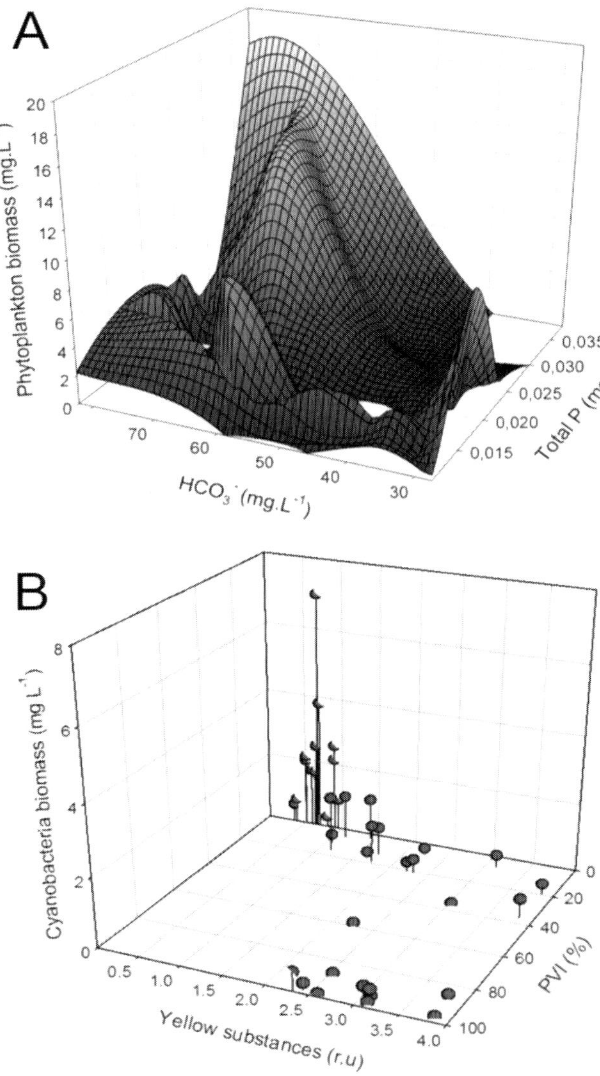

Figure 5. (A) Phytoplankton biomass (mg L^{-1}) in relation to total phosphorus (TP) (mg L^{-1}) and bicarbonate HCO_3^- (mg L^{-1}); (B) Relationship between cyanobacterial biomass (mg L^{-1}) and concentration of yellow substances (r.u.) and PVI (%) (Finkler Ferreira, 2009).

Cyanobacteria were already identified as being sensible organisms to humic substances exposure (Sun et al., 2005; Sun et al., 2006). Their much simpler cellular ultrastructure may justify their failure compared to coccal green algae success in humic substances enriched ecosystems (Prokhotskaya and Steinberg, 2007). It was also stressed that polyphenols, originated from lignin and cellulose decomposition of vascular plants, are able to complex chemically with many enzymes, particularly with alkaline phosphatases produced by phytoplankton and bacteria in order to hydrolyze dissolved organic phosphorous compounds

(Stewert and Wetzel, 1982b). This complexation affects free enzymes as well as those bound to cell surfaces. Because of the inactivation of the phosphatase activity cells may experience phosphorus limitation and consequently are induced to expend greater amounts of energy to synthesize more enzymes (Wetzel, 1992).

An important aspect is whether macrophyte density must attain a certain threshold before they have a significant effect on phytoplankton community structure and biomass (Sondergaard and Moss, 1998). Schriver et al. (1995) reported a threshold level of 15-20% PVI above which a sudden decrease of 60% in the biovolume of phytoplankton occurred. Van den Berg et al. (1998) found the water to be clearer at a highly vegetated bed of *Chara aspera* (100%) compared with a non-vegetated area of the same lake. In that study, it was found that the water contained less chlorophyll *a*, detritus and inorganic suspended solids within the beds of *C. aspera*. At macrophyte coverages from 16-50%, there were already clear evidences of reduction in density and biomass of algae, as well as shifts in the phytoplankton community. Moreover, it was observed that flagellates were likely to occur within the beds because they could escape from sedimentation, while outside there was higher density of cyanobacteria and green algae (van den Berg et al., 1998).

In this survey, Finkler Ferreira (2009) observed that the influence of the submerged vegetation was beyond the limits of the vegetation stand. It depended on the wind intensity and direction, which determined the distribution of humic substances and other chemical components. Thus, the effects of macrophytes on water physical-chemical features and phytoplankton community were perceived over the edge of the stands (~1000m), further out the interface zone. Although few studies have also demonstrated phytoplankton growth inhibition in situ, this theory still needs more investigation. Considering this possibility, the allelopathic activity of the main submerged macrophytes from Lake Mangueira was tested in microcosms against a toxic cyanobacterial strain *Microcystis aeruginosa* (Finkler Ferreira, 2009).

Laboratory Assesment: Coexistence Experiments with Submerged Macrophytes and Cyanobacteria

Coexistence experiments in batch cultures are a suitable way to explore allelopathy and resource competition, and thereby to perceive the real negative effects of macrophytes on phytoplankton, or vice-versa (Lürling et al., 2006). The use of small aquaria provides an opportunity for direct evaluation of donor-target species interactions and minimizes light competition. In addition, one can include different factors (e.g., light and nutrients), which are important variables for gathering evidence on the modes of action of allelopathy or competition. With the evaluation of phosphorus concentration in short term experiments, though which allelopathy trials are often performed, the determination of net foliar uptake by the plants could contribute to understand the magnitude and importance of this mechanism (Lombardo and Cooke, 2003). Because nutrient determinations (especially orthophosphate) in batch media are seldom addressed directly, these measurements, coupled with the analysis of growth inhibition of phytoplankton, may shed light on the search for competitive and allelopathic macrophyte species (Gross et al., 2007).

In this approach, the potential allelopathic effects of the macrophytes *Cabomba caroliniana, Egeria densa*, *Ceratophyllum demersum* and *Myriophyllum spicatum*, inhabitants of subtropical and tropical ecosystems, were tested against a toxic strain of the cyanobacterium *Microcystis aeruginosa* (Finkler Ferreira, 2009).

Material and Methods

The coexistence experiments were carried out in small aquaria in order to test the effect of the physical presence of macrophytes under three levels of light intensity and orthophosphate concentrations (Finkler Ferreira, 2009). The three levels of light intensity were 40-60, 120-140 and 220-240 µmol PAR (quanta m^{-2} s^{-1}), chosen to represent low, medium and high luminosity, respectively. The three concentrations of PO_4^{-3}-P were 0.25, 0.5, and 1 mg L^{-1}, chosen to represent eutrophic conditions at which phosphorus would not be limiting within 7 days of batch culture. In addition to standard monitoring of growth and photosynthesis, measurements of orthophosphate were carried out to address the occurrence and strength of competitive-like interspecific interactions. Based on the experimental outcome, we aimed at determining the nutrient and light thresholds above which allelopathy or competition are likely to inhibit the growth of this cyanobacterium.

The submerged macrophyte species were collected from Lake Mangueira, along with local sand to be used as s substrate for the plants. The aquaria were inoculated with 50 µg L^{-1} of chlorophyll *a* (~3.0 × 10^4 particles mL^{-1}) of exponentially growing *M. aeruginosa*. Three replicate aquaria were used for controls and treatments. The positive controls were plant-free aquaria containing only the modified ASM-1 medium and *M. aeruginosa*. In order to evaluate the possible shading effect, plastic *C. demersum*-like plants were added in the same weight (10 g fresh weight) as real macrophytes that caused similar light reduction. In total, the experimental design comprised 54 aquaria with 2 L of medium each, divided into groups with 18 aquaria for each factor (Figure 7).

The aquaria were incubated for a week (7-8 days) under controlled climate conditions of temperature (23-24 °C) and photoperiod (12:12 h). The cyanobacteria growth was monitored daily for chlorophyll *a* (chla), photosystem II efficiency (ΦPSII), and algal density in each aquarium. The chla content (µg L^{-1}) and the ΦPSII efficiency were determined using the PHYTO-PAM® phytoplankton analyzer (Walz, Germany). The density of cells was estimated using the CASY® cell counter (Innovatis, Germany).

Summary of Results

Under the three different light regimes, the presence of macrophytes clearly affected the growth of *M. aeruginosa* over the incubation period, represented by the chlorophyll *a* concentration in the treatments. This effect caused a significant difference in chla concentrations among the treatments under 60 ($F(6,14) = 4.556$; $P < 0.01$), 120 ($F(6,14) = 5.438$; $P < 0.001$), and 240 µmol PAR ($F(6,14) = 6.647$; $P < 0.001$). Similarly, the macrophytes caused a reduction in cyanobacteria density compared to the control, being statistically significant for the lowest and highest light intensity ($F(6,14) = 3.075$, $P < 0.05$; $F(6,14) = 5.387$, $P < 0.001$, respectively). At the lowest PO_4^{-3}-P concentration, the density of *M. aeruginosa* was significantly attenuated by the four macrophyte species. However, only *M. spicatum* and *C. caroliniana* reduced the chla and density values significantly at all PO_4^{-3}-P concentrations (Figure 6). The mean chla based growth rates (not shown) of *M. aeruginosa* were negative in the presence of the macrophytes *C. caroliniana* and *M. spicatum* at all PO_4^{-3}-P concentrations, while the other macrophytes did not promote the same effect (Figure 7).

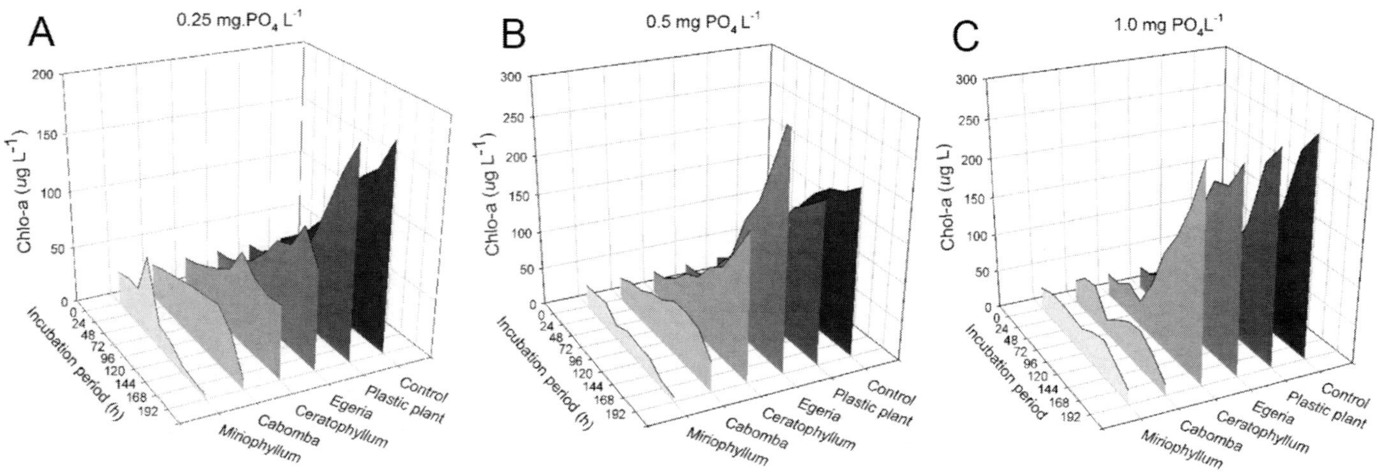

Figure 6. Chlorophyll *a* concentrations of *Microcystis aeruginosa* in the presence of 4 species of macrophytes and in 2 controls at three PO_4^{-3} concentrations: 0.25 (A), 0.5 (B), and 1.0 mg L^{-1} (C) (Finkler Ferreira, 2009).

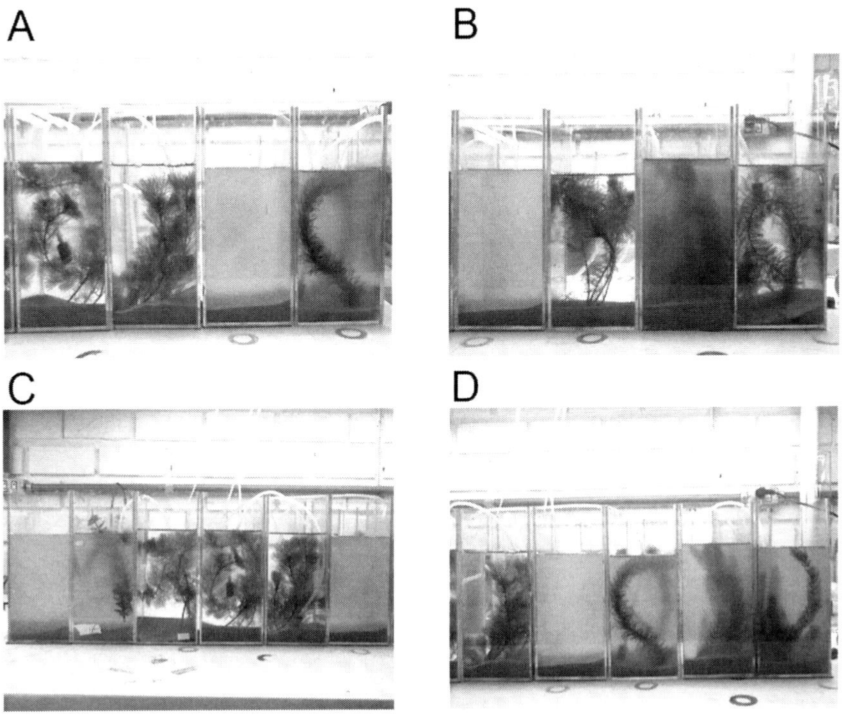

Figure 7. Evidence of allelopathy after 7 days of incubation: chl *a* reduction in the treatments with the macrophytes *Cabomba caroliniana* (A, plants on the left and C) and *Myriophyllum spicatum* (B) compared with the controls. In contrast, *Ceratophyllum demersum* (B, plants in the third aquarium on the right and D, plants on the right) and *Egeria densa* (A and D) did not cause the same effect (Finkler Ferreira, 2009).

Discussion

The significant reduction in growth of *M. aeruginosa* in the presence of *M. spicatum* and *C. caroliniana* indicated the allelopathic potential of these macrophytes. For *C. caroliniana*, there is only one previous report of its allelopathy on phytoplankton growth (Nakai et al., 1999). In their study, *C. caroliniana* negatively affected *Microcystis aeruginosa* and *Anabaena flos-aquae*. Additionally, *E. densa* was also found to have a significant negative effect on both cyanobacteria.

Our findings differ from these previous results probably because those studies were based on testing plant extracts on algae cultures in vitro. This type of approach, although useful for preliminary tests, does not always accurately represent the possible effects of real interactions (Gross et al., 2007). In summary, *E. densa* did not have a direct effect in controlling the cyanobacterium *M. aeruginosa* through allelopathy. Certainly, the role of *Egeria densa* as an allelopathic macrophyte cannot be disconsidered at all based only on *M. aeruginosa* and hence is still not clear.

With respect to *C. demersum*, this species has been extensively mentioned as having allelopathic effects against phytoplankton and epiphyton (Van donk and Gullati, 1995). Wium-Andersen et al. (1983) isolated a sulfur compound with allelopathic properties from *C. demersum*. Jasser (1995) concluded that the release of organic compounds contributed to a decline of cyanobacteria by changing the phytoplankton structure. However, our results did not corroborate such findings. It is likely that allelochemicals possibly produced by this species are not sufficiently effective against *M. aeruginosa*. Perhaps further experiments testing other cyanobacteria species might shed light on potential differences in sensitivities among target species.

The allelopathic effects of *M. spicatum* have been investigated by other researchers (Gross et al., 1996; Leu et al., 2002; Nakai et al., 1999; Körner and Nicklisch, 2002), including the isolation and identification of the polyphenolic acid tellimagrandin II, which showed inhibitory effects on cyanobacterial growth (Gross et al., 1996; Leu et al., 2002). These compounds had a much smaller effect on green algae and diatoms, which, in fact, is a remarkable finding as all phytoplankton species could be considered competitors with submerged macrophytes. Nonetheless, this difference in sensitivity of different taxonomic groups to allelopathic substances is likely to influence the competitive balance between cyanobacteria and other algae (Van Donk and van den Bund, 2002).

It was also stressed that polyphenols, originating from lignin and cellulose decomposition of vascular plants, are able to complex chemically with many enzymes, particularly alkaline phosphatases produced by phytoplankton and bacteria to hydrolyze dissolved organic phosphorus compounds (Stewert and Wetzel, 1982b). This complexation occurs with free enzymes as well as with those bound to cell surfaces. Because of the inactivation of the phosphatase activity, the cells may experience phosphorus limitation and consequently may be induced to expend greater amounts of energy in order to synthesize more enzymes (Wetzel, 1992). At the end of the experiment, *M. spicatum* and *C. caroliniana* released large amounts of humic substances (> 10 mg L^{-1} of dissolved carbon), which colored the water after 7 days of incubation. The release of humic substances is an indication that chemicals are being released into the water by the plants. However, whether allelochemicals are released through continuous metabolic excretion, or only after contact with phytoplankton, is still an important subject to be evaluated (Hilt et al., 2006). Furthermore, the effect of humic substances, in general, on phytoplankton is a rather contradictory matter. Some studies have observed cyanobacteria to be sensitive to exposure to humic acids (Sun et al., 2005; Sun et al., 2006; Steinberg et al., 2006; Prokhotskaya and Steinberg, 2007; Steinberg et al., 2008). Their much simpler cellular ultrastructure would explain their failure compared to the success of coccal green algae in humic-enriched ecosystems (Prokhotskaya and Steinberg, 2007). In contrast, Kosakowska et al., (2007) found that filtered humic substances could enhance the growth of *Microcystis* sp., even under low iron concentrations.

Conclusion

Both field and experimental approaches provided clear indications that submerged macrophytes exerted different potential mechanisms to inhibit phytoplankton and also cyanobacterial growth. In Lake Mangueira, the phytoplankton community composition, especially cyanobacteria, was influenced either through humic acids and/or allelopathic compounds present in the water nearby the macrophyte submerged stands (Finkler Ferreira, 2009).

In order to elucidate this hypothesis, we carried out coexistence experiments with the main macrophyte species from Lake Mangueira. The experiments allowed us to confirm that macrophytes such as *Myriophylum spicatum* and *Cabomba caroliniana* were able to inhibit the growth of *M. aeruginosa* through allelopathy. The introduction of macrophytes such as *C. caroliniana* and *M. spicatum* could be used to mitigate cyanobacterial blooms at chlorophyll *a* concentrations of 50 µg L^{-1}. On the other hand *Egeria densa* and *Ceratophyllum demersum* showed to be important nutrient competitors, being able to deplete orthophosphate from the water at high rates. Among the nutrient competitive species, *C. demersum* showed the highest PO_4^{-3}-P uptake rates. Together with *E.densa*, it could be used to stabilize the clear water vegetated state at meso to oligotrophic conditions in water bodies under restoration process.

In summary, both through allelopathy and nutrient competition, the submerged macrophytes may be considered natural mitigators to be applied for management, restoration and conservation of subtropical shallow lakes.

HETEROTROPHIC BACTERIA

One of the paradigms of the Aquatic Microbial Ecology is that heterotrophic bacterial communities are tightly associated with phytoplankton primary production, but in the last years this paradigm has been under debate (Fouilland and Mostajir, 2010). Previous reports considered that phytoplankton was responsible for driving bacterial metabolism and diversity (Horner-Devine et al., 2003) and also synchronize bacterioplankton dynamics with climatological events (Kent et al., 2007). Fouilland and Mostajir's work (2010) showed in a comprehensive cross ecosystem comparison that in most cases phytoplankton primary production is not enough to cope with bacterioplankton's total carbon demand (production + respiration). In this case, alternative primary producers like macrophytes can provide either directly or indirectly (via support of epiphytes) higher amounts of DOC than algae (Wetzel, 1992). Even though the macrophyte-derived carbon is proportionally more refractory than the carbon derived from algae, the dependence of bacteria on this carbon source could be explained by the larger amounts released by macrophytes (Wetzel, 1992).

Several studies have demonstrated that the relationship between macrophytes and bacterioplankton is not always positive for the last though. Macrophytes can liberate compounds that inhibit bacterial growth, either directly or indirectly, the last through primary inhibition of phytoplankton (Scheffer and Jeppesen, 1997; Körner and Nicklisch, 2002), through increased refractoriness of leachates after photochemical transformation (Farjalla et al., 2001) or still by complexation of polyphenols released by the plants with bacterial enzymes, which inactivates them (Wetzel, 1991).

Submerged macrophytes have been related to the production of allelopathic compounds, and the inhibitory effects on the growth of algae is widely known and reported (Körner and Nicklisch, 2002; van Donk; van de Bund, 2002; Leu et al., 2002; Gross et al., 2007; Mulderij et al., 2007). Inhibitory effects directly on bacteria have been hypothesized by field observations of lower bacterial abundance, metabolism and diversity in areas under the influence of mixed macrophyte stands when compared to areas without their influence (Rooney and Kalff, 2003; Huss and Wehr, 2004; Wu et al., 2007; They et al., 2010; They et al., 2013a; They et al., 2013b). Again, good candidates could be exudated polyphenolic compounds (Wetzel, 1991; Boavida and Wetzel, 1998). Hence, global effects of macrophytes can be controversial, as macrophytes at one side favor bacteria by liberating organic substrates and nutrients but at another side compounds capable of inactivate their enzymes. However, *in situ* and experimental evidence for this mechanistic explanation is still poorly elucidated.

In Brazil, the coastal plain of the Rio Grande do Sul state presents several water bodies with a great variety of macrophytes. Among these, *Potamogeton pectinatus* is a submersed macrophyte originated from temperate regions and found in streams, lakes and coastal areas with low salinity (Wijk, 1989). This plant presents long and narrow leaves, with thin and perennial stems and rhizomes. It shows apical development, with great production of stratified biomass under favorable conditions (Colares et al., 2007). Compounds with algicidal activity have been isolated from this species (Waridel et al., 2003) and it has been used in investigative studies of allelopathic effects on phytoplankton species (Körner and Nicklisch, 2002). Ethanolic extract from a related species (*Potamogeton perfoliatus*) has been found to have biocidal effect on protozoans by inhibition of an enzyme responsible for fatty acid biosynthesis (Orhan et al., 2006). Sequential extractions with light petroleum, ethyl acetate and methanol and further separation by HPCL of another *Potamogeton* species, *P. natans*, have rendered six furano-*ent*-labdanes, six γ-lactone-labdanes and one *nor*-diterpene with deleterious effects on phytoplankton and zooplankton species.

P. pectinatus occurs in the estuarine region of Patos Lagoon (southern Brazil), environment where it has been already reported that the bacteria can be more dependent on macrophytes than algae as a source of carbon (see review in Cesar and Abreu, 2001). If bacteria depend largely on macrophytes as a carbon source in this environment, and the last also present an inhibitory effect on the former, this ecological interaction becomes complex with possible repercussions on all trophic web. Hence, this study aimed at: I) assessing density, biovolume and biomass of bacteria, flagellates and ciliates and also a set of environmental variables and exoenzymatic activity in areas under or not the influence of this macrophyte in sequential water bodies; II) testing the effect of fractions of *P. pectinatus* (extract, decomposing leafs and stems and also the surrounding water of stands), assessed by means of growth inhibition of the components of the microbial food web (bacteria, flagellates and ciliates) and exoenzymatic activity through an experimental approach. Our hypotheses were that there was lower density, biovolume and biomass of the components of the microbial food web and lower exoenzymatic activity in areas where this macrophyte is present and that the fractions of the plant are able to induce lower densitiy, biovolume and biomass of microbial food web components, as well as lower exoenzymatic activity.

MATERIAL AND METHODS

Sampling Areas

P. pectinatus was collected in the Bolaxa stream, which is located in the coastal plain of Rio Grande do Sul state (southernmost state in Brazil) and oriented longitudinally in the NE-SW direction, following the beach ridges that are characteristic of the region. This stream belongs to the city of Rio Grande, and is still under low antropic influence, presenting only punctual loadings of livestock wastes (Carvalho and Silva, 1998). With 4 km of extension, the stream is densely colonized by floating and submersed macrophytes, and is part of the Environmental Protection Area of the lake Verde (Green Lake), in which it disembogues and through it links to the Patos Lagoon (Carvalho and Silva, 1998). In the present study the Bolaxa stream was the sampling point for the macrophyte and stands' surrounding water collection, and also the point under the influence of the macrophyte (hereafter called point 1) (Figure 8).

The Lagoa Verde presents an extension of 2 km (Figure 8) and communicates to the estuary of Patos Lagoon via Saco da Mangueira, through which it receives contribution of salt water. In the confluence of these two water bodies we collected samples for the area not under the influence of the macrophytes (hereafter point 2). The variation in the salinity in this lake depends on the regime of rain and wind (Costa et al., 1988; Colares et al., 2007). The estuarine region of the Patos Lagoon comprises about 10% of the total area of the Lagoon - 971 km^2) and the only communication to the sea is through a narrow channel 4 Km long and 0.5-3 km wide (Chao et al., 1985). At this point was proceeded the collection of the inoculums and the incubation water for the experiment. The idea of using this water was to submit a bacterial community not adapted to the phytochemicals of the plant to an experimental stimulation of exponential growth in the presence of the macrophyte fractions and test for an inhibitory effect on their growth. The ecological meaning of this is that the estuarine bacterial community eventually enters the lake during some periods of the year and is exposed to the compounds released by the plants then.

Field Sampling and Experimental Deign

At the point 01 it was collected stems and leaves of *P. pectinatus* in plastic bags *ziploc®* containing water from the point. For the stands surrounding water, the water was collected in a 10 L plastic container. Samples for bacteria, flagellates and ciliates were collected in amber glass bottles of 10 and 50 mL of capacity, respectively.

For the analyses of nutrients, 250 mL of water was collected in plastic bottles. It was also collected samples or the analysis of absorbance at low- (250 nm) and high molecular weight (365 nm) substances, colored dissolved organic carbon (430 nm), turbidity (750 nm) and phosphatase activity in glass amber bottles previously baked in oven for 3 h at 450°C. All samples were taken in triplicates. Sample for enzymes had to be frozen or less than a month (-20°C), even though not recommended (Chróst and Velimirov, 1991).

For the point 02 we collected samples for bacteria, flagellates and ciliates, nutrients and absorbance analyses the same way as for point 01.

The water employed for the experiment was sampled in the point 03 after pre-filtration in 200 μm zooplankton net. The water was kept in two plastic containers of 10 L at room temperature until the experimental set up (< 24 h). It was also collected samples for bacteria, flagellates, ciliates and absorbance analyses the same way as for points 01 and 02.

In the laboratory, the water from point 03 was filtered in glass fiber filters (0.4 μm of mean retention) and distributed in 12 erlenmeyers of 1 L of capacity, and autoclaved at 121°C for 30 minutes (incubation water). The experiment was carried with three treatments and one control (C). The first treatment (T1) consisted of the addition of a methanolic extract of the vegetal material from *P. pectinatus*. The second treatment consisted of the addition of surface sterilized vegetal material, and the third treatment (T3) consisted of the addition of the stands surrounding water. In all treatments and also in the control it was added an inoculum containing the water from the point 03 without filtration and autoclaving (only field pre-filtration).

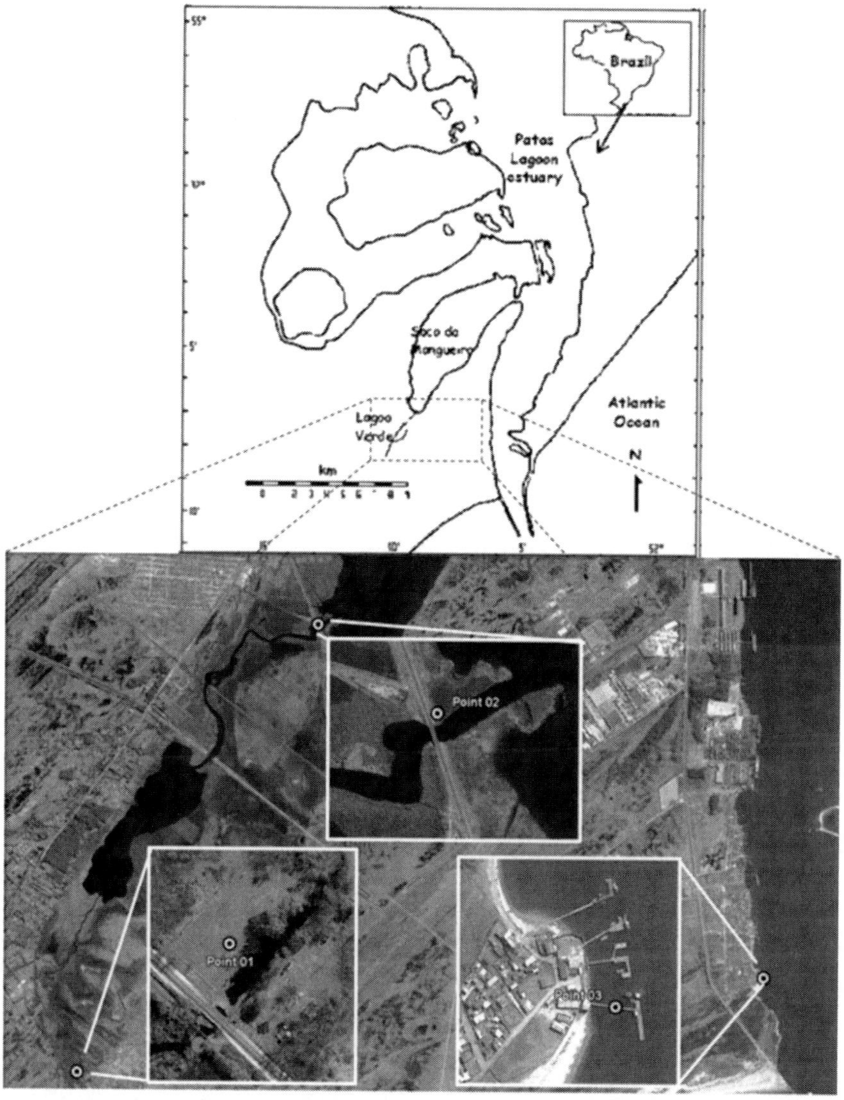

Figure 8. Sampling points: Points 01 (Bolaxa Stream – area with *P. pectinatus*), 02 (confluence of *Lagoa Verde* and *Saco da Mangueira* – control area without *P. pectinatus*) and 03 (estuarine zone, where the inoculum and incubation water were taken) (modified from Colares et al., 2007 and Google Earth, 2010). Sampling and experiment were carried in May 2010.

For T1, it was employed 700 mL of incubation water + 35 mL of methanolic extract + 50 mL of inoculum. This extract was obtained after the modification of the method of Erhard and Gross (2006): 80 mg of plant leaves and stems per mL of solvent were immerged in a 50% aqueous methanol (Merck) solution for 2 h twice, with a total volume of 100 mL of extract. The extraction was carried under agitation with magnetic stirrer bar and at room temperature. After that, the extract was pre-filtered with glass fiber filters (GF – 0.4 μm of mean retention) and then filter sterilized in polycarbonate membrane of 0,2 μm of mean retention.

For T2, we employed 750 mL of incubation water + 5g of surface sterilized vegetal material + 50 mL of inoculum. The vegetal material was sterilized following the method described by Kane and Gilman (1991) and by Gross et al., (1996). The plant was first washed thoroughly with tap water and the excess water removed. Then, it was cut in small parts containing leaves and stems and weighted 5g for each triplicate. Each batch was treated with an immersion in 200 mL of a sodium hypochlorite 1% and Tween 20 0.01% for 12 minutes, followed by three rinses with autoclaved Milli-Q water.

For T3 we employed 700 mL of incubation water + 50 mL of stands surrounding water + 50 mL of inoculum. The surrounding stand water was filtered sterilized in polycarbonate membrane of 0.2 μm of porosity.

For C we employed 750 mL of incubation water and 50 mL of inoculum.

The experimental flasks were incubated for 5 days at 20°C with a photoperiod of 12/12 h. We collected samples at the beginning and 24, 48, 72 and 120 h - 50 mL aliquots for flagellates and ciliates, 10 mL for bacteria and also 30 mL for absorbance analyses and enzymatic activity assays (frozen) in amber glass flasks. All samples were taken the same way as for the field samples.

Bacterial Density, Biovolume and Biomass

The samples were fixed in formaldehyde (4% final concentration) and kept under refrigeration. For the slides mounting, 2 mL of the water was filtered in polycarbonate membrane of 0.2 μm of porosity previously stained with Irgalan Black. The filters were stained with an acridine orange solution 1% (w/v) for approximately 10 minutes and then destained with pre-filtered (0.2 μm) destilled water. Slides were mounted with mineral oil Nujol and analyzed under epifluorescence microscope Zeiss (light filters: BP 450-490; FT 510; LT 520). A total of 15 images were captured per filter and then processed through the software Image Tool (3.0) for countings and cellular dimensions measurements (modified after Kepner and Pratt, 1994).

The bacterial biovolume was estimated through the major axis measure and approximation to a spherical solid through the equation I and the biomass estimated using the allometric conversion factor of Norland (1993).

$$\text{Biovolume}\ (\mu m^3) = \frac{4}{3}\pi r^3 \qquad \text{(Equation I)}$$

where r is the estimated radius (μm) of the spherical solid approximation.

Flagellates and Ciliates Density, Biovolume and Biomass

Samples were preserved in lugol 2% and kept in the dark. The density of flagellates and ciliates was estimated through countings of 2.1 and 10 mL, respectively, using sedimentation chambers under inverted microscope Zeiss (Utermöhl, 1958). The major axis of the organisms was measured and for the biovolume estimate it was also used the approximation to a spherical geometric solid (Sun and Liu, 2003). The biomass was estimated after the biovolume with the empirical conversion factors of Riemann et al. (1990): 0.12 pg C μm^{-3} for flagellates and 0.19 pg C μm^{-3} for ciliates (Putt and Stoecker, 1989).

Relative Concentration of Labile and Refractory Substances, DOC_{430} and Humic Substances

The relative concentration estimate of labile substances, refractory substances, DOC-specific light absorption (DOC_{430}) (Stepanauskas et al., 2000) and turbidity were estimated with absorbance at 250, 365, 430, and 750 nm, respectively. The ratio 250:365 nm (humic substances ratio, HS ratio) was also employed to estimate the proportion of low and high weight compounds, respectively, being an indication of the general refractoriness state of the organic substances in the water (Strome and Miller, 1978; Lindell et al., 1995).

Enzymatic Assays

Phosphatase activity was assessed fluorometrically (TD 700 - Turner Designs) with the MUF-phosphate (4-methylumbelliferyl-phosphate) analog approach (Chróst 1989; Chróst 1991; Hoppe, 1993). In summary, 2.7 mL of each sample (in triplicates) was added to a test tube-like cuvette + 0.3 mL (100 μM of final concentration, above saturation) of MUF-phosphate (Sigma-Aldrich). The essay was carried for 4 h in the dark at 20°C and readings were done under 300-400 nm excitation and 410-610 nm emission after addition of 0.3 mL of ammonium hydroxide-glycin buffer pH 10.5 in order to achieve maximum emission of MUF (Chróst and Krambeck, 1986). Killed controls were carried for initial incubation and control for non-enzymatic hydrolysis by boiling the samples for 20 minutes in waterbath (Hoppe, 1983). All essays were carried in laminar-flow hood and with filter-barrier tips in order to minimize contamination.

Nutrients

Total nitrogen (TN) and phosphorus (TP) were analyzed spectrophotometrically according to modification of the methods described by Strickland and Parsons (1972) after digestion with potassium persulfate (Carmouze, 1994).

Statistical Analysis

Field environmental variables (bacterial, flagellates and ciliates density, biovolume and biomass, absorbances at 250, 365, 430, 750 and 250:365 nm ratio, total nitrogen, total phosphorus and enzymatic activity) were tested for differences among the inoculum, point 01 and point 02 by Kruskal-Wallis and Mann-Whitney *a posteriori* tests, as the variables did not match assumptions for parametric statistics. The tests were performed in PAST 1.9 (Hammer et al., 2001).

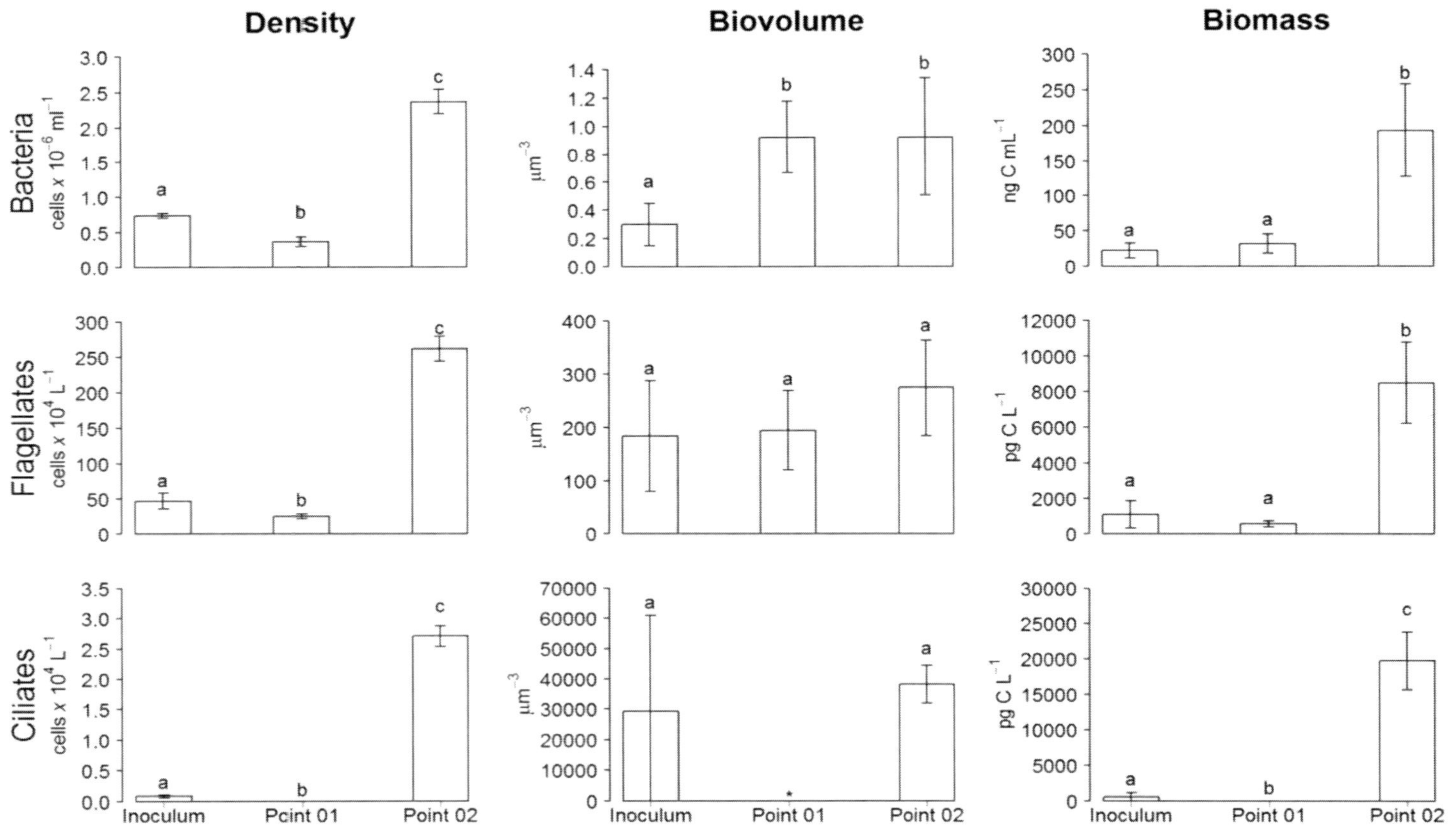

Figure 9. Density, mean cellular biovolume and total biomass of bacteria, flagellates and ciliates in the inoculum and sampling points 01 (with *P. pectinatus*) and 02 (without *P. pectinatus*). Same small letters indicate that the medians do not differ. (*) indicate that no ciliate was recorded and no comparison could be made.

Results

The Table 4 summarizes the general characteristics of the sampling points. In general, the point 02 (control point, without P. pectinatus) presented the highest bacterial, flagellates and ciliates densities and biomasses while point 01 (where P. pectinatus was present) presented the lowest ones (ciliates, however, were not detected in point 01). The biovolume was very similar among the points, except for bacteria, which showed lower biovolumes in the inoculum compared to the points (Figure 9).

Table 4. Location and environmental variables of sampling points and inoculums

	Location	Coordinates	pH	Salinty
Inoculum	Patos Lagoon Estuary	-32° 09' 03.09" -52° 06' 04.45”	8.0	30
Point 01	Bolaxa stream	-32° 09' 35.0" -52° 11' 16.8"	7.4	0
Point 02	Confluence of Lagoa Verde and Saco da Mangueira	-32° 06' 50.4" -52° 09' 28.0"	8.1	2

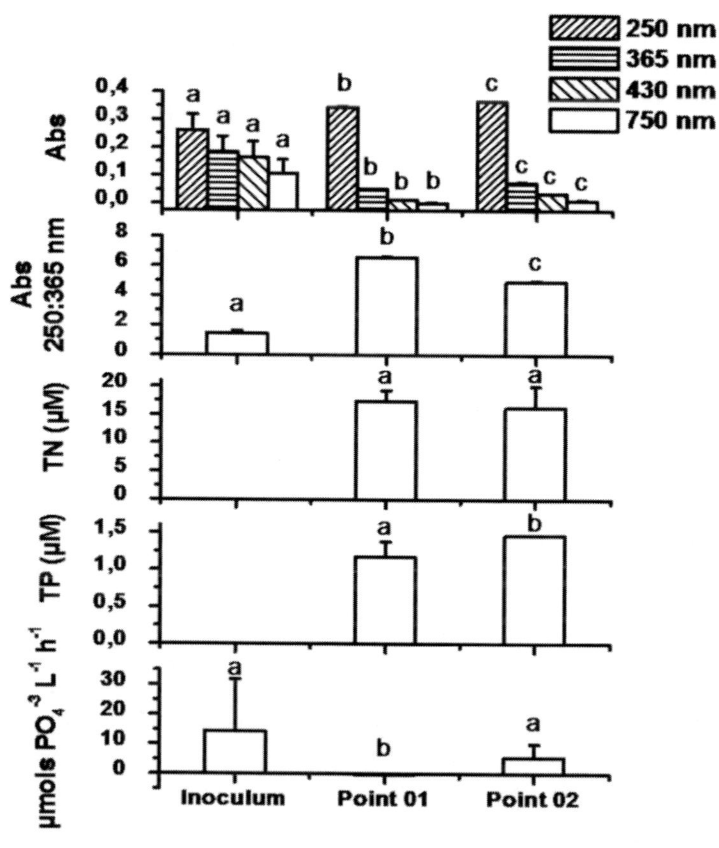

Figure 10. Low- (Abs 250 nm) and high-molecular-weight substances (Abs 365 nm), DOC_{430} (Abs 430 nm), turbidity (Abs 750 nm), humic substances ratio (Abs 250:365 nm), total nitrogen (NT), total phosphorus (PT) and phophatase activity in the inoculum and sampling points 01 (with *P. pectinatus*) and 02 (without *P. pectinatus*). Data for TN and TP in the inoculum not available.

Low-molecular-weight substances (LMW, Abs 250 nm) were higher in point 02, followed by point 01 and inoculum. High-molecular-weight substances (HMW, Abs 365 nm), and turbidity were higher in the inoculum, followed by point 02 and point 01. DOC_{430} was higher in the inoculum, followed by point 01 and point 02. The HS ratio was higher in point 01, followed by point 02 and then the inoculum. For nutrients, TN was very similar between the two points and the phosphorus was lower in the point 01. Phosphatase activity was higher in the inoculum, but not different from point 02; in the point 01 it was undetectable though (Figure 10).

Experiment

For LMW substances, T1 (methanolic extract) and T2 (vegetal material) presented initial values higher than T3 (stands´ surrounding water) and C (control). Between 0-24 h, these values remained stable for T1, T3 and C, increasing considerably for T2 until 24 h. T3 and C showed always a very similar behavior (Figure 11).

HMW substances were also initially higher for T1 and T2 (T1 > T2). Between 0-24 h, T1, T3 and C decreased, while for T2 there was a steep increase until 72 h, reaching higher values than T1. From 24-72 h there was an increase for all treatments, and after this period they remained relatively stable, with exception of T2, which increased slightly. As for LMW substances, T3 and C displayed a similar behavior (Figure 11)

DOC_{430} showed initial values: T1 > T2 > T3 ~ C and decreased in all treatments between 0-24 h. From 24-72 h there was an increase in all treatments, which stabilized after 72 h. At the end, the pairs T1/T2 and T3/C reached similar values (Figure 11).

Turbidity was very similar for all treatments, with an initial drop until 24 h, increasing between 24-72 h and relatively stability until the end of the experiment (Figure 11).

Regarding the HS ratio, it was mostly in the descending order: T2 > T1 > T3 > C. In all treatments this ratio increased until 24 h (decreasing in the refractability) and then tended to decrease in all treatments until 72 h, remaining stable after this period (Figure 11).

Total bacterial density was much lower than the initial inoculum density at the beginning of the experiment. Between 0-24 h all treatments presented similar behavior with a slight decrease in density. T1 presented an exponential growth between 24-48 h and after 72 h approached the plateau. T2 presented density close to zero until 48 h, period after it started exponential growth until the end of the treatment, overcoming values of T1. T3 and C had exponential growth between 24-48 h and stabilized after 72 h (Figure 12).

Mean bacterial cell biovolume was small ($< 2\ \mu m^3$) and similar for all treatments in the period 0-24 h. Between 24-48 h there was an exponential increase, in descending order C > T3 > T1 > T2, respectively. Between 48-72 h the treatments T1, T2 and T3 presented increase, while C showed a drop in biovolume. After 72 h all treatments had a drop, reaching similar values at the end of the experiment (Figure 12).

Initial bacterial biomass was low (close to zero), being not altered until 24 h. Between 24-48 h the treatments T1, T3 and C increased (highest for T1). T2 remained very close to zero. Between 48-72 h T1, T3 and C kept increasing, with similar values for T3 and C. T2 had a steep increase, approaching T1. After 72 h, T1, T3 and C started to decrease and T2 stabilized with a higher value than T1 (Figure 12).

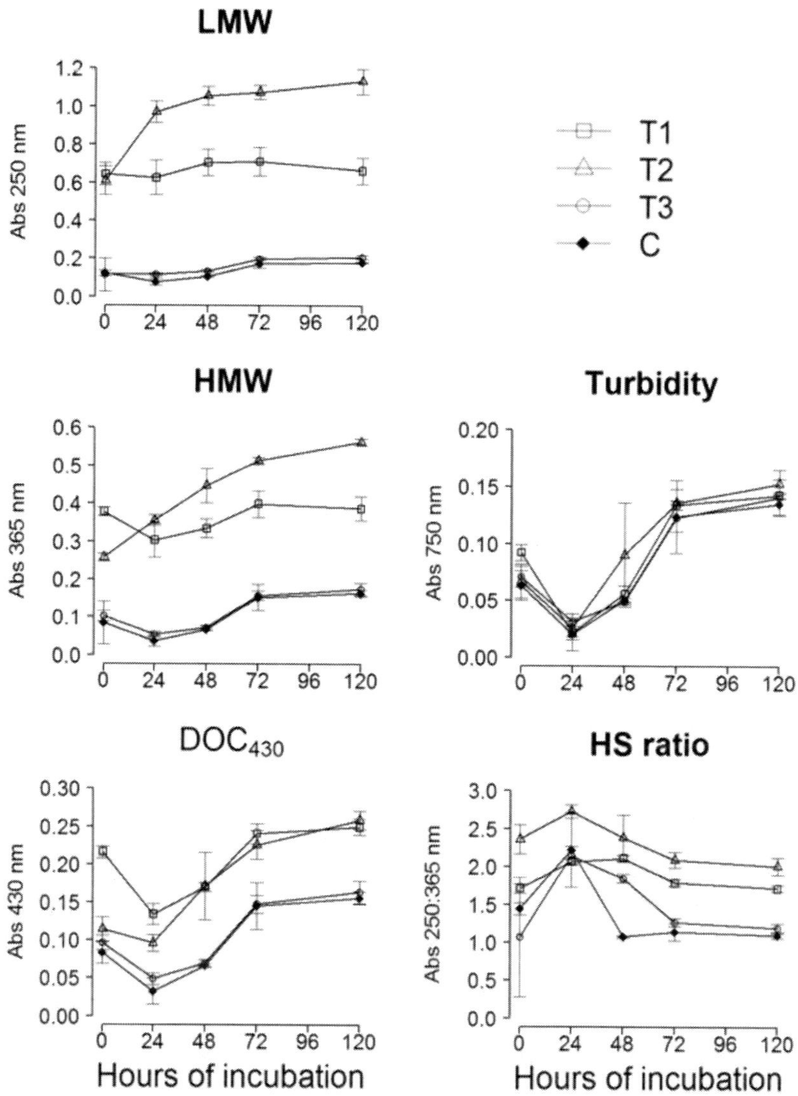

Figure 11. Evolution along the experiment of the concentration of LMW substances (Abs 250 nm), HMW substances (Abs 365 nm), DOC_{430} (Abs 430 nm), turbidity (Abs 750 nm) and humic substances ratio (Abs 250:365 nm) for the treatments T1 (methanolic extract), T2 (vegetal material), T3 (stands' surrounding water) and C (control).

Flagellates density showed a distinct behavior among the treatments for the period 0-24 h; T1 presented a slight increase, while T2 remained stable, T3 decreased slightly and C had a steep drop, remaining then practically without alteration until 120 h. Between 24-48 h, T1 had a steep drop until values close to zero, increasing slightly between 72-120 h. T2 started an exponential growth which peaked at 72 h, decreasing at 120 h. T3 remained stable until 72 h, increasing slightly at 120 h (Figure 12).

Mean flagellate cellular biovolume decreased between 0-24 h in all treatments, more accentuated for T1. Between 24-72 h there was an increase in the biovolume for all treatments, except for C, which decreased from 48-72 h. After 72 h, T1 and T2 showed decrease in biovolume, while T3 and C remained relatively stable (Figure 12).

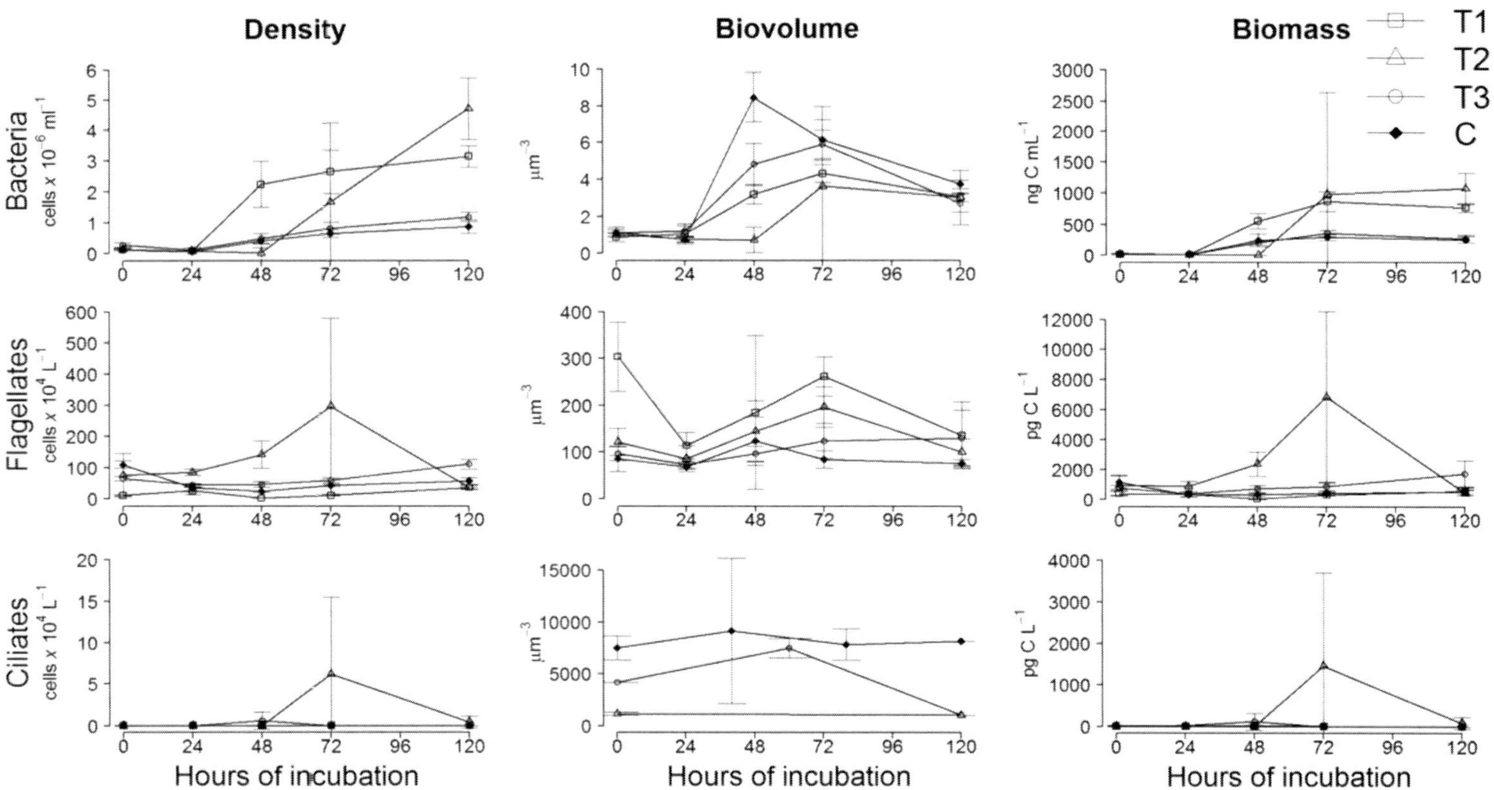

Figure 12. Evolution along the experiment of bacteria, flagellates and ciliates density, biovolume and biomass for the treatments T1 (methanolic extract), T2 (vegetal material), T3 (stands surrounding water) and C (control).

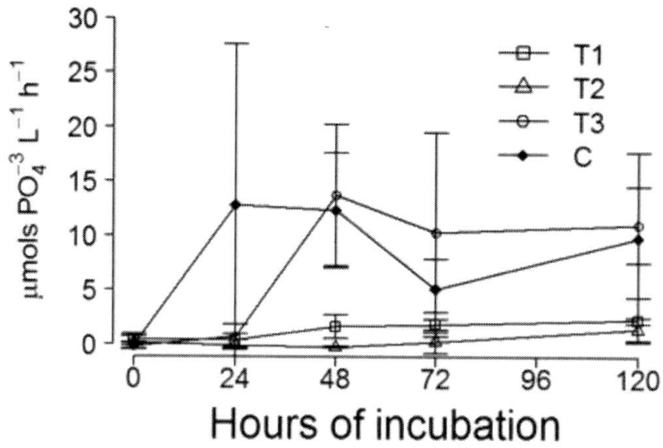

Figure 13. Phosphatase activity (total enzymatic hdrolysis of MUF-phosphate) in the treatments T1 (methanolic extract), T2 (vegetal material), T3 (stands´ surrounding water) and C (control).

Flagellates biomass was stable and close to zero for T1, only presenting increase between 48-72 h. T2 also initiated stable, showing its higher phase of growth between 24-72 h, dropping then steeply until the end of the treatment, when it approached values close to T1 and C.T3 increased slightly until the end of the experiment, with higher values than T1, T2 and C. C, at its turn was stable during all the incubation period (Figure 12).

Ciliates density and biomass exhibited the same pattern. No ciliates were ever recorded in T1. For all the other treatments, the ciliates density between 0-48 h was practically zero, except for T3 which showed a discrete growth after 24 h and then drop to zero in the end of the incubation. T2 showed a steep growth between 48-72 h (peak), and also decreased to zero at 120 h. C presented only values close to zero, but were recorded during all the incubation period (Figure 12). For T2, ciliates appeared only after 72 h, period after that their mean cell biovolume remained low (< 1000 μm^3) and without alteration until the end of the experiment. C and T3 presented higher values of cellular volume (C > T3). T3 showed an increase until 24 h and then a drop after 48 h, period after that no more ciliates were recorded for this treatment. For C, there was a slight increase in biovolume between 24-48 h and a slight drop between 24-72 h, not being recorded after that (Figure 12).

Rates of enzymatic hydrolysis of phosphate were always lower for T1 and T2 (always T1 > T2), but they tended to slightly increase until 120 h. C showed an exponential increase between 0-24 h, remaining relatively stable between 24-48 h, dropping between 48-72 h and then increasing again at 120 h. T3 at its turn, showed a similar behavior but with a 24 h lag; it peaked at 48 h, decreased slightly until 72 h and then remained stable thereafter (Figure 13).

Discussion

Overall we have found conflicting results between field and experimental data. Despite lower bacterial and flagellate abundance and no ciliates in locations under the influence of the macrophyte *P. pectinatus*, experimental addition of material originated from the plant stimulated the inhibiting bacterial growth in T1, but at the same time induced a one day longer lag phase in treatment with the addition of decomposing stems and leaves. This apparent paradox, however, reveals one of the possible mechanistic explanations by which the plants affect negatively heterotrophic bacteria.

With regard to the points, the slightly lower total phosphorus concentration in the point where the macrophyte *P. pectinatus* was present suggests that there is a higher consumption of this nutrient by the plants given that in general macrophytes are good competitors for nutrients, as observed and suggested in other works (Barko and James, 1998; Rooney and Kalff, 2003; Huss and Wehr, 2004). Contrasting to density, bacterial and flagellates biovolume was very similar in points 01 and 02; this could be explained by the low competition of the few organisms that can thrive in this environment. What is striking is that no ciliates were recorded; given the low predator pressure, either flagellates or bacteria should be benefitted. Ciliates absence on one way could be explained by the low density of prey (bacteria and flagellates), but on the other side, studies in temperate lakes have shown that one of the impacts of macrophytes on trophic structure can be lower concentrations of ciliates and flagellates in macrophyte beds, attributed to higher grazing by macrozooplankton hiding in the vegetation in order to avoid fish predation (Jürgens and Jeppesen, 1997; Jeppesen et al., 2002). For the experiment there was an initial growth in bacterial abundance for all treatments but T2 after 24 h, as well as of mean biovolume, with the highest values always observed in C. The highest biovolumes recorded for T3 and C in relation to T1 and T2 could be explained by the lower biovolumes of flagellates and ciliates, assuming a preference of these organisms for larger bacterial cells. The peak of bacterial biovolume, however, arose first in C at 48 h, being dislodged for all other treatments at 72 h, showing a close relationship with the peak of flagellates biovolume. Ciliates in T3 and C were less abundant, which could have led to error in density and consequently biovolume and biomass estimation. Hence, it was not possible to conclude about these organisms in these treatments.

The effect of methanol (solvent control) was not tested, but Erhard and Gross (2006) did not find any inhibitory effect on phytoplankton or cyanobacteria species. For the treatment with methanolic extract (T1), there was not bacterial growth inhibition, as it was the treatment with highest initial growth rate, being later overcome by the treatment with decomposing vegetal material (T2). This high initial growth rates could be explained by the higher presence of LMW substances (e.g., simple sugars) that are co-extracted in the extract preparation (Gross et al., 2007). If there are actually present inhibitory compounds, they might have been masqueraded. This holds true in the light of other studies that have successfully inhibited phytoplankton and cyanobacteria growth with methanolic extracts (Gross et al., 1996; Erhard and Gross 2006; Mulderij et al., 2007).

Flagellates and ciliates in T1 presented a slight increase until 24 h, but a drastic decrease in biovolume. This could be due to a drop in bacterial density during this period. Hence, these small flagellates that increased can be auto-/mixotrophic (not discriminated in function of the method employed). The increase in biovolume in the period 24-72 h was probably in response to the increase in bacterial biovolume. Density, however, did not follow the increase in bacterial biovolume. Ciliates, however, never occurred in this treatment, suggesting that the extract could have had an efficient inhibitory effect on these organisms. The extraction process also gave T1 initial higher amounts of LMW and HMW substances, and consequently higher carbon content in relation to the other treatments. This could be related to the higher rates of bacterial initial growth of this treatment.

For T2, the longer lag in bacterial growth until 48 h suggests that growth could have been prevented either by an inhibitory effect or by higher bacterivory. The higher growth rates than

T1 after 48 h could be explained by the degradation of the compounds responsible for this effect, once there was an increase in content of LMW, HMW substances and DOC_{430} during the experiment. The growth after the lag could be expected once there was not continuous addition of these compounds (Gross et al., 2007). According to Nakai et al. (1999), initial or quasi-continuous addition of allelochemicals may underestimate an inhibitory effect, whether it does exist, when compared to coexistence experiments. Some fractions of extracts from *Stratiotes aloides* were also shown to stimulate growth of the phytoplankton species under K and P limitation, as the extract was probably used as a source of these nutrients (Mulderij et al., 2007), which according to the authors masqueraded the results.

Yet for T2, there was not a clear explanation for the flagellate growth in the period 0-24 h. This could be also due to the presence of small auto-/mixotrophic flagellates, once flagellate biovolume also decreased in this period. The peak of density and biovolume followed immediately after the peak in bacteria density and biovolume at 48 h, indicating grazing. Ciliates also grew in response to bacterial and flagellate growth.

T3 and C presented the lowest amounts of LMW and HMW substances by the absence or much lower initial input of plant material than T1 and T2. This confirms the hypothesis that the concentration of substances in T3 was very similar to C. The DOC_{430} was slightly higher in T3 than C, but it assumed a similar behavior to C after 48 h. HS ratio was higher for T3 and C than other treatments during all period, but possibly because of the high amounts of LMW substances in T1 and T2. The very similar behavior of T3 and C could have two explanations: by the possible low concentration of compounds in T3 given the low volume added or by the degree of degradation of the compounds, once they are found in several states of degradation in the water that surrounds the macrophyte stands. Once it was found that the compounds can have a degradation time of 48 h (as observed by the beginning of exponential growth of T2) and that a considerable amount of time elapsed between collection of the water and it being used in the experiment (< 24 h), there may have occurred a significant degradation of these compounds during this period. Erhard and Gross (2006) also found that a period of 24 h incubation with exudated allelochemicals from *Elodea nuttalii* caused a weaker inhibitory effect on the growth of *Anabaena* than a 4 h exposition.

The release of allelopathic compounds can be an effective competitive trait of submerged macrophytes (Mulderij et al., 2007) and the extraction method used in the present study has been shown to successfully extract allelopathically active compounds (most probably polyphenols) by previous works. There has been a debate about the nature of the inhibitory compounds; Basile et al., (1999) have demonstrated that flavonoids derived from mosses exhibit antibacterial activity against several species of Gram negative bacteria, but not against those Gram positive tested. However, Erhard and Gross (2006) did not find any allelopathic activity of purified *Elodea* flavonoids on *Anabaena* (Cyanobacteria, Gram negative), which leaves the question open of which plant compound/fraction could be the allelopathically active on bacteria. Mulderij et al. (2007) have found that allelopathically active compounds from *Stratiotes aloides* are moderately lipophilic and not likely to be phenolic compounds. One important possibility is the synergistic effects that have been observed by Nakai et al. (2000), who found growth inhibition effect of polyphenols from the submerged macrophyte *Myriophyllum spicatum* in isolation and in combination on *Microcystis* aeruginosa. *P. pectinatus*, however, did not show any effect on chlorophyll *a* concentration and photosystem II inhibition on the cyanobacteria *Aphanizomenon flos-aquae* in coexisting dialysis membrane experiments, while other submersed macrophytes *Myriophillum spicatum* and *Ceratophyllum demersum* did (Körner and Nicklisch, 2002). They have also found a pattern of several species of phytoplankton chlorophyll *a* and photosystem II activity initial inhibition, as well as later effects too.

To our knowledge, field studies investigating direct allelopathic effects on heterotrophic bacteria have not been carried and our observations showed that bacteria, flagellates and ciliates are negatively affected in the presence of the plant, from a density and enzymatic activity point of view. Why is bacterial density so low in the field even though flagellates and ciliates density is also very low and bacteria do can grow much more in plant's extract? The answer could be an indirect effect caused by macrophytes: inhibition of enzyme activity. Despite the fact that substrate utilization rates in natural waters are frequently extremely low (Stumm-Zollinger, 1966), our observations showed that enzyme activity is much lower in areas under the influence of the plants and in the presence of methanolic extracts and the decomposing stems and leaves from *P. pectinatus*.

There is an important distinction between the mechanisms through lower enzymatic activity occurs; enzymes may be either inhibited (enzymes are produced, but remain inactive) or repressed (enzymes production is suppressed by the substrate it produces - negative feedback) (Stumm-Zollinger, 1966). Inhibition can be caused by complexation with polyphenols (Wetzel, 1991). It has been experimentally demonstrated that humic acids can bind to alkaline phosphatases, inactivating them, a process that can be reversed by UV irradiance. One of the ecological implications of this process is that these complexes act as storages of these enzymes, with important impacts on the global aquatic metabolism and nutrient cycling (Boavida and Wetzel, 1998). Decreases in alkaline phosphatase activity associated with the presence of polyphenols in crude extract and in isolation from the submersed macrophyte *Myriophyllum spicatum* has already been demonstrated (Gross et al., 1996). Repression at the other side, can be caused by low concentration of substrate (German et al., 2011) and/or high concentrations of the own enzyme´s end products. It is well established that the last can have a negative feedback on the enzyme synthesis (Datta, 1969) and Stumm-Zollinger (1966) found that in mixed bacterial populations growing on sugars the presence of glucose and absence of inorganic nitrogen sources represses the expression of phenylalanine deaminase, resulting that this aminoacid is not utilized in the presence of glucose. Increasingly lower substrates concentration, at its turn, represses enzyme synthesis, as it becomes energetically too costly (German et al., 2011). Utilization of phosphate compounds was not assessed in this work, and whether any substrate can repress/inhibit the uptake of phosphate remains unanswered.

Two possible global scenarios can be drawn from this work: I) in the field, *P. pectinatus* continuously releases compounds (polyphenolic compounds are good candidates) that bind to the enzymes making them unfunctional, which prevents bacterial growth and hence their predator´s growth too. In the experimental approach, the methanolic extract co-extracted labile substances (which do not require enzymatic processing before consumption), which then was not able to cause inhibition. In the decomposing plant treatment, there is an initial and limited input of polyphenols, which prevented growth for one day longer, but subsequent leaching of leaves liberated labile compounds, supporting then bacterial growth. Limited input of polyphenols was also overcome by bacterial enzyme production. This could be supported if a high ratio of living:non-living plant biomass is present in the field. II) a second explanation could be higher predation pressure in the presence of the plant, as observed by the strong cascading effects of flagellates on bacterial density and biovolume increase (selecting larger cells, similar to what was observed in the field); this could only be supported by field observation if the few bacterial predators present can be somehow more effective in predation and respond rapidly to pulses of bacterial growth; lower enzymatic activity could be then explained by negative feedback (repression) triggered by release of labile substances by the plant.

The overall ecological implications of these findings are that in places where these potentially allelopathic macrophytes occur, there is lower bacterial density and biomass, implying a lower bacterial metabolism, which draws attention to important and not yet accounted ecological implications. This is a rather novel topic in Aquatic Microbial Ecology, and future efforts are necessary to identify the ecosystems where this complex interactions takes place and to which extent they impact energy and matter cycling in aquatic ecosystems.

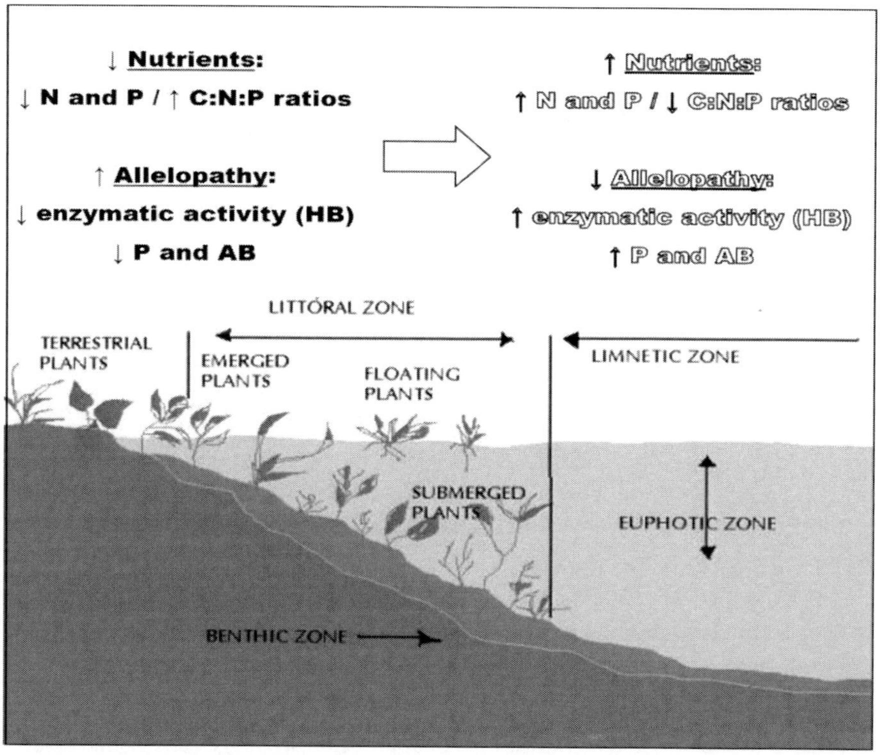

Figure 14. Conceptual model of the allelopathic effect of macrophytes in subtropical shallow lakes. HB, heterotrophic bacteria; P, phytoplankton; AB, autotrophic bacteria; C, carbon; N, nitrogen; P, phosphorus. Modified from: http://wgbis.ces.iisc.ernet.in/energy/water/paper/ETR25/ecosystem.htm.

CONCLUSION

The three studies presented along this chapter provided evidence that macrophytes in subtropical shallow lakes are not only strong competitors for nutrients modifying the chemical and physical environment in the areas they colonize (Figure 14), but also present allelopathic effects with important repercussions on the trophic web. They are able to impact negatively autotrophic - through reduced phytoplankton biomass, diversity and species richness, and particularly cyanobacterial biomass - as well as heterotrophic - through reduced bacterial, flagellates and ciliates density, biovolume and biomass and enzymatic activity - organisms in these lakes.

Although the effects on autotrophic organisms have received a lot of attention in the past, allelopathic effects of macrophytes on heterotrophic bacteria and other components of the microbial food webs have been completely neglected. The evidence of lower enzymatic activity associated with the plants or their extracts presented in this study provides a mechanistic explanation for the lower bacterial abundance and metabolism repeatedly found in areas colonized by macrophytes in these lakes and reinforces our hypothesis that the microbial loop functions poorer under the influence of these plants.

ACKNOWLEDGMENTS

This work was supported by the Brazilian National Council for Scientific and Technological Development-CNPq/Long Term Ecological Research Network of Brazil, Site 7: Sistema Hidrológico do Taim. We also acknowledge the assistance provided by the National Institute of Environment (IBAMA, ESEC TAIM).

REFERENCES

Barko, J. W.; James, W. F. Effects of submerged aquatic macrophytes on nutrient dynamics, sedimentation and ressuspension. *Ecol. Stu.* 1998, 131, 197-214.

Basile, A.; Girdano, S.; López-Sáez, J. A.; Cobianchi. R. C. Antibacterial activity of pure flavonoids isolated from mosses. *Phytochemistry*. 1999, 52, 1479-1482.

Blindow, I.; Hootsmans, M. J. M. Allelopathic effects from *Chara* spp. on two species of unicellular green algae. In: Hootsmans, M. J. M.; Vermaat, J. E., editors. *Macrophytes: A Key to Understanding Changes Caused by Eutrophication in Shallow Freshwater Systems*. Delft: IHE Report Series; 1991; v. 21; 139–144.

Blindow, I.; Anderson, G.; Hargeby, A.; Hansson, S. Long term pattern for alternative stable states in two shallow eutrophic lakes. *Freshwater Biol.* 1993, 30, 159-167.

Boavida, M.-J.; Wetzel, R. G. Inhibition of phosphatase activity by dissolved humic substances and hydrolytic reactivation by natural ultraviolet light. *Freshwater Biol.* 1998, 40, 285-293.

Burks, R. L.; Jeppesen, E.; Lodge, D. M. Macrophyte and fish chemicals suppress *Daphnia* growth and alter life-history traits. *Oikos* 2000, 88, 139–148.

Burns, C. W.; Dodds, A. Food limitation, predation and allelopathy in a population of *Daphnia carinata*. *Hydrobiologia* 1999, 400, 41–53.

Canfield, D. E.; Shireman Jr. J.V.; Colle, D. E.; Haller, W. T.; Watkins, I. I. C. E.; Maceina, M. J. Prediction of chlorophyll *a* concentrations in Florida lakes: importance of aquatic macrophytes. *Can. J. Aquat. Sci.* 1984, 41, 497–501.

Carmouze, J. P. *O metabolismo dos ecossistemas aquáticos. - Fundamentos teóricos, métodos de estudo e análises químicas*. São Paulo: FAPESP; 1994.

Carpenter, S. Submersed vegetation: an internal factor in lake ecosystem succession. *Am. Nat.* 1981, 118, 372-383.

Carvalho, R.V.; Silva, K.G. Caracterização Ambiental do Sistema Arroio-Lagoa do Bolaxa. Uma Futura Área de Preservação Ambiental. *Anais da XI Semana Nacional de Oceanografia*. 1998, 46-48.

Cesar, D. E.; Abreu. P. C. Ecology of microorganims in southern Brazil: state of art. In: Faria, B. M.; Farjalla, V. F.; Esteves, F. A., editors. *Aquatic Microbial Ecology in Brazil*. Rio de Janeiro: Series Oecologia Brasiliensis. 2001; v. 9; 153-172.

Chao, L. H.; Le Pereira; Vieira, J. P. Estuarine fish community of the dos Patos Lagoon, Brazil. A baseline study. In: Yanez-Arancibia, A., editor. *Fish Community Ecology in Estuaries and Coastal Lagoons: Towards an Ecoystem Integration*. Mexico: UNAM Press. 1985; 429-450.

Chróst, R. J. Characterization and significance of β-glucosidase activity in lake water. *Limnol. Oceanogr*. 1989, 34, 660-672.

Chróst, R. J.; Krambeck, H. J. Fluorescence correction for measurements of enzyme activity in natural waters using methylumbelliferyl-substrates. *Arch. Hydrobiol.* 1986, 106, 79-90.

Chróst, R. J. Environmental control of the synthesis and activity of aquatic microbial ectoenzymes. In: Chróst, R. J., editor. *Microbial Enzymes in Aquatic Environments*. New York: Springer-Verlag. 1991; 29-59.

Chróst, R. J.; Velimirov, B. Measurement of enzyme kinetics in water samples: effect of freezing and soluble stabilizer. *Mar. Ecol. Prog. Ser.* 1991, 70, 93-100.

Colares, I. G.; Schlee, M. D. B.; Santos, L. C.; Magalhães, U. A. S. Variação da biomassa e produtividade de *Potamogeton pectinatus* L. (Potamogetonaceae) na Lagoa Verde, Rio Grande, RS. *Iheringia Sér. Bot*. 2007, 62, 131-137.

Costa, C. S. B.; Seeliger, U.; Kinas, P. G. The effect of wind velocity and diretion on the salinity regime in the lower Patos Lagoon estuary. *Ciência e Cultura.* 1988, 40, 909-912.

Datta, P. Regulation of branched biosynthetic pathways in bacteria. *Science.* 1969, 165, 556-562.

DellaGreca, M.; Fiorentino, A.; Isidori, M. Bioactive Compounds from Potamogetonaceae on Aquatic Organisms. In: Macias, F. A.; Galindo, J. C. G., Molinillo, J. M. G.; Cutler, H. G. *Allelopathy: Chemistry and Mode of Action of Allelochemicals*. London: CRC Press. 2003; 35-56.

Dorgelo, J.; Heykoop, M. Avoidance of macrophytes by *Daphnia longispina. Verh. Int. Verein. Limnol*. 1985, 22, 3369–3372.

Elakovich, S. D.; Wooten, J. W. Allelopathic, herbaceous, vascular hydrophytes. *Allelopathy-organisms, processes, and applications*. ACS Symposium series. Washington: American Chemical Society. 1995; 582; 58–73.

Erhard, D.; Gross, E. M. Allelopathic activity of Elodea Canadensis and Elodea nuttalli against epiphytes and phytoplankton. *Aquat. Bot.* 2006, 85, 203-211.

Farjalla V. F.; Anesio, A. M.; Bertilsson, S.; Granéli. W. Photochemical reactivity of aquatic macrophyte leachates: abiotic transformations and bacterial response. *Aquat. Microb. Ecol.* 2001, 24, 187-195.

Finkler Ferreira, T. The role of submerged macrophytes on the dynamics of shallow subtropical lakes: the importance for restoration and conservation of Lake Mangueira, Southern Brazil [PhD thesis]. Porto Alegre: Universidade. Federal do Rio Grande do Sul; 2009.

Fouilland E.; Mostajir B. Revisited phytoplankton carbon dependency of heterotrophic bacteria in freshwaters, transitional, coastal and oceanic waters. *FEMS Microbiol. Ecol.* 2010, 73, 419-429.

Garcia, A. M.; Vieira, J. P.; Bemvenuti, M. A.; Motta Marques, D. M. L.; Burns, M.; Moresco, A., Condini, V. Checklist comparison and dominance patterns of the fauna at Taim Wetland, South Brazil. *Neotrop. Icththyol.* 2006, 4, 261-268.

Gazulha, V.; Montú, M.; Marques, D. M.; Bonecker, C. C. Effects of natural banks of free-floating plants on zooplankton community in a shallow subtropical lake in Southern Brazil. Braz. *Arch. Biol. Technol.* 2011, 54, 745-754.

German, D. P.; Chacon, S. S.; Allison, S. D. Substrate concentration and enzyme allocation can affect rates of microbial decomposition. *Ecology.* 2011, 92, 1471-1480.

Google Earth (v. 5.2.1.1329, beta) [software]. Mountain View, CA: Google Inc.; 2010.

Gopal, B.; Goel, U. Competition and allelopathy in aquatic plant communities. *Bot. Rev.* 1993. 59, 155–210.

Gross, E. M. Allelopathy in benthic and littoral areas: case studies on allelochemicals from benthic cyanobacteria and submerged macrophytes. In: Inderjit Dakshini, K. M. M.; Foy, C. L., editors. *Principles and Practices in Plant Ecology: Allelochemical Interactions.* Boca Raton: CRC Press; 1999.

Gross, E. M.; Meyer, H. O; Schiling, G. Release and ecological impact of algicidal hdrolysable polyphenols in *Myriophyllum spicatum. Phytochemistry.* 1996, 41, 133-138.

Gross E. M. Differential response of tellimagrandin II and total bioactive hydrolysable tannins in an aquatic angiosperm to changes in light and nitrogen. *Oikos*, 2003, 103, 497-504.

Gross, E. M.; Erhard, D.; Ivanyi, E. Allelopathic activity of *Ceratophyllum demersum* L. and *Najas marina* ssp. intermedia. *Hydrobiologia* 2003, 506, 583–589.

Gross, E. M.; Hilt (nee Körner), S.; Lombardo, P.; Mulderij, G. Searching for allelopathic effects of submerged macrophytes on phytoplankton – state of the art and open questions. *Hydrobiologia.* 2007, 584, 77-88.

Hammer, Ø.; Harper, D. A. T.; Ryan, P. D. PAST: PAlaeontological Statistics software package for education and data analysis. *Palaeontol. Electron.* 2001, 4, 1-9.

Hilt S.; Ghobrial, M.G. N.; Gross, E. M. *In situ* allelopathic potential of *Myriophyllum verticillatum* (Haloragaceae) against selected phytoplankton species. *J. Phycol.* 2006, 42, 1189–1198.

Hootsmans, M. J. M.; Blindow, I. Allelopathic limitation of algal growth by macrophytes. In: W. van Viersen; Hootsmans M. J. M.; Vermaat, J. E., editors. *Lake Veluwe, Dynamics of a macrophyte dominated system under eutrophication stress.* Netherlands: Kluwer Academic Publishers. 1994; 175-192.

Hoppe, H.-G. Significance of exoenzymatic activities in the ecology of brackish water: measurements by means of methylumbelliferyl-substrates. *Mar. Ecol. Progr. Ser.* 1983, 11, 299-308.

Hoppe, H.-G. Use of fluorogenic model substrates for extracellular enzyme activity (EEA) measurement of bacteria. In: Kemp, P. F.; Sherr, B. F.; Sherr, E. B.; Cole, J. J., editors. *Handbook of Methods in Aquatic Microbial Ecology.* Boca Raton: Lewis. 1993; 423-431.

Horner-Devine, M. C.; Leibold, M. A.; Smith, V. H.; Bohannan, B. J. M. Bacterial diversity patterns along a gradient of primary productivity. *Ecol. Lett.* 2003, 6, 613-622.

Horppila, J.; Nurminen, L. Effect of submerged macrophytes on sediment resuspension and internal loading in Lake Hiidenvesi (southern Finland). *Wat. Res.* 2003, 37, 4468–4474.

Horppila, J.; Nurminen, L. Effect of different macrophyte growth forms on sediment and P resuspension in a shallow lake. *Hydrobiologia* 2005, 545, 167–175.

Hulot, F. D.; Huisman, J. Allelopathic interactions between phytoplankton species: the roles of heterotrophic bacteria and mixing intensity. *Limnol. Oceanogr.* 2004, 49, 1424-1434.

Huss, A. A.; Wehr, J. D. Strong indirect effects of a submersed aquatic macrophyte, *Vallisneria Americana*, on bacterioplankton densities in a mesotrophic lake. *Microb. Ecol.* 2004, 45, 305-315.

Jasser, I. The influence of macrophytes on a phytoplankton community in experimental conditions. *Hydrobiologia.* 1995, 306, 21–32.

Jeppesen, E.; Lauridsen, T. L.; Kairesalo, T.; Perrow, M. R. Impact of submerged macrophytes on fish–zooplankton interactions in lakes. In: Jeppesen, E.; Søndergaard, M.; Christoffersen, K., editors. *The Structuring Role of Submerged Macrophytes in Lakes*. New York: Springer. 1998; 91–114.

Jeppesen, E.; Søndergaard, M.; Søndergaard, M.; Christoffersen, K.; Theil-Nielsen, J.; Jürgens, K. Cascading trophic interactions in the littoral zone: an enclosure experiment in shallow lake Stigsholm, Denmark. Arch. Hydrobiol. 2002, 153, 533-555.

Jürgens, K.; E. Jeppesen, Cascading effects on microbial food web structure in a dense macrophyte canopy. In: Jeppesen, E.; Søndergaard, Ma.; Søndergaard, Mo.; Christoffersen, K., editors. *The structuring role of submerged macrophytes in lakes.* Netherlands: Springer. 1997; v. 131; 262-274.

Kane, M. E.; Gilman, E. F. *In vitro* propagation and bioassay systems for evaluating growth regulator effects on *Myriophillum* species. J. Aquat. Plant Manage. 1991, 29, 29-32.

Kent, A. D.; Yannarell, A. C.; Rusak, J. A.; Triplett, E. W.; McMahon, K. D. Synchrony in aquatic microbial community dynamics. *ISME J.* 2007, 1, 8-47.

Kepner, R. L. J. R.; Pratt, J. R. Use of fluorochromes for direct enumeration of total bacteria in environmental samples: past and present. *Microbiol. Rev.* 1994, 58, 603-615.

Kluijver, A.; Ning, J.; Liu, Z.; Jeppesen, E.; Gulati, R. D.; Middelburg, J. J. Macrophytes and periphyton carbon subsidies to bacterioplankton and zooplankton in a shallow eutrophic lake in tropical China. *Limnol. Oceanogr.* 2015, 60, 375–385.

Körner, S.; Nicklisch, A. Allelopathic growth inhibition of selected phytoplankton species by submerged macrophytes. J. Phycol. 2002, 38, 862-871.

Kosakowska A.; Nedzi, M.; Pempkowiak, J. Responses of the toxic cyanobacterium *Microcystis aeruginosa* to iron and humic substances. *Plant Physiol. Bioch.* 2007, 45, 365-370.

Kufel L.; Kufel I. Chara beds as nutrient sinks in shallow lakes- a review. *Aquat. Bot.* 2002, 72, 249-260.

Lammens, E. H. R. R. Causes and consequences of the success of bream in Dutch eutrophic lakes. *Hydrobiol. Bull.* 1989, 23, 11–18.

Lauridsen, T. L.; Lodge, D. M. Avoidance of *Daphnia magna* by fish and macrophytes: chemical cues and predator-mediated use of macrophyte habitat. *Limnol Oceanogr.* 1996, 41, 794–798.

Lauridsen, T. L.; Pedersen, L. J.; Jeppesen, E.; Søndergaard, M. The importance of macrophyte bed size for cladoceran composition and horizontal migration in a shallow lake. *J. Plankton Res.* 1996, 18, 2283–2294.

Leu, A.; Krieger-Liszkay, A.; Goussias, C.; Gross, E. M. Polyphenolic allelochemicals from the aquatic angiosperm *Myriophyllum spicatum* inhibit photosystem II. *Plant Physiol.* 2002, 130, 2011-2018.

Lindell, M. J.; Granéli, W.; Tranvik, L. J. Enhanced bacterial growth in response to photochemical transformation of dissolved organic matter. *Limnol. Oceanogr.* 1995, 40, 195-199.

Lloyd, R. J. M.; Ghelardi, A. Table for Calculating the `Equitability' Component of Species Diversity. *J. Anim. Ecol.* 1964, 33, 217-225.

Lombardo, P.; Cooke, G. D. *Ceratophyllum demersum* – phosphorus interactions in nutrient enriched aquaria. *Hydrobiologia.* 2003, 497, 79-90.

Lund, J. W. G.; Kipling, C.; Le Cren, E. D. The inverted microscope method of estimating algal numbers and the statistical basis of estimations by counting. *Hydrobiologia.* 1958, 11, 143-170.

Lurling, M.; van Geest, G.; Scheffer, M.. Importance of nutrient competition and allelopathic effects in suppression of the green alga *Scenedesmus obliquus* by the macrophytes *Chara*, *Elodea* and *Myriophyllum*. *Hydrobiologia.* 2006, 556, 209-220.

Meerhoff, M.; Iglesias, C.; Teixeira-de-Mello, F.; Clemente, J. M.; Jensen, E.; Lauridsen L.; Jeppesen, E. Effects of habitat complexity on community structure and predator avoidance behaviour of littoral zooplankton in temperate versus subtropical shallow lakes. *Freshwater Biol.* 2007, 52, 1009–1021.

Mjelde, M.; Faafeng, B. *Ceratophyllum demersum* (L.) hampers phytoplankton development in some small Norwegian lakes over a wide range of phosphorus level and geographic latitude. *Freshwater Biol.* 1997, 37, 355–365.

Molisch, H. Der Einflusseiner Pflanze auf die andere-Allelopathie. Jena: Fischer. 1937.

Moss, B. Engineering and biological approaches to the restoration from eutrophication of shallow lakes in which aquatic plant communities are important components. In: Gulati, R. D.; Lammens, E. H. R. R.; Meijer, M.-L.; van Donk, E. *Biomanipulation. Tool for Water Management.* London: Kluwer Academic Press. 1990; v. 61; 367-377.

Mulderij, G.; van Nes, E. H.; Donk, E. V. Macrophyte-phytoplankton interactions: the relative importance of allelopathy versus other factors. *Ecol. Model.* 2007, 204, 85-92.

Nakai, S.; Inoue, Y.; Hosomi, M.; Murakami, A. Growth inhibition of blue-green algae by allelopathic effects of macrophytes. *Water Sci. Technol.* 1999, 39, 47-53.

Nakai, S.; Inoue, Y.; Hosomi, M.; Murakami, A. *Myriophyllum spicatum*-released allelopathic polyphenols inhibiting growth of blue-green algae *Microcystis aeruginosa*. *Wat. Res.* 2000 34, 3026-3032.

Norland, S. The relationship between biomass and volume of bacteria. In: Kemp, P. F.; Sherr, B. F.; Sherr, E. B.; Cole, J. J., editors. *Handbook of Methods in Aquatic Microbial Ecology.* USA: Lewis Publishers. 1993; 339-345

Orhan, I.; Sener, B.; Atici, T.; Brun, R.; Perozzo, R.; Tasdemir, D. Turkish freshwater and marine macrophyte extracts show in vitro antiprotozoal activity and inhibit FabI, a key enzyme of *Plasmodium falciparum* fatty acid biosynthesis. *Phytomedicine.* 2006, 13, 388-393.

Padisák, J.; Crossetti, L. O.; Naselli-Flores, L Use and misuse in the application of the phytoplankton functional classification: a critical review with updates. *Hydrobiologia.* 2009, 621, 1-19.

Persson, L.; Johansson, L.; Andersson, G.; Diehl, S.; Hamrin, S.F. Density dependent interactions in lake ecosystems whole lake perturbation experiments. *Oikos* 1993, 66, 193–208.

Persson, L.; Eklov, P. Prey refuges affecting interactions between piscivorous perch and juvenile perch and roach. *Ecology* 1995, 76, 70–81.

Prokhotskaya V.Y.; C. E. W., Steinberg. Differential sensitivity of a coccal green algal and a cyanobacterial species to dissolved natural organic matter (NOM). *Environ. Sci. Pollut. R.* 2007, 14: 11-18.

Putt, M.; Stoecker, D. K. An experimentally determined carbon:volume ratio for marine "oligotrichous" ciliates from estuarine and coastal waters. *Limnol. Oceanogr.* 1989, 34, 1097-1103.

Reigosa, M. J.; Sanchez-Moreiras, A.; Gonzalez, L. Ecophysiological approach in allelopathy. CRC Cr. Rev. Plant Sci. 1999, 18, 577–608.

Reynolds, C.S. Vegetation in the pelagic: a model for ecosystem theory. In: O. Kinne, editor. Germany: Ecology Institute; Excellence in Ecology Series. 1997.

Reynolds, C.S.; Huszar, V.; Kruk, C.; Naselli-Flores, L.; Melo, S. Towards a functional classification of the freshwater phytoplankton. *J. Plankton Res.* 2002, 24, 417-428.

Riemann, B.; Sørensen, H. M.; Bjørnsen, P. K.; Horsted, S. J.; Jensen, L. M.; Nielsen, T. G.; Søndergaard, M. Carbon budgets of the microbial food web in estuarine enclosures. *Mar. Ecol. Progr. Ser.* 1990, 65, 159-170.

Rodrigues, L. R.; Fontoura, N. F.; Motta Marques, D. Food-web structure in a subtropical coastal lake: how phylogenetic constraints may affect species linkages. *Mar. Freshwater Res.* 2013, 453-465.

Rooney, N.; Kalff, J. Submerged macrophyte-bed effects on water-column phosphorus, chlorophyll *a*, and bacterial production. *Ecosystems*. 2003, 6, 797-807.

Scheffer M. Ecology of Shallow Lakes. London: Chapman and Hall. 1998.

Scheffer, M.; Hosper, S. H.; Meijer, M. L.; Moss, B.; Jeppesen, E. Alternative equilibria in shallow lakes. *Trends Ecol. Evol.* 1993, 8, 275–279.

Scheffer, M.; Jeppesen, E. Alternative stable states. In: Jeppesen, E.; Søndergaard, M.; Søndergaard, M.; Christoffersen, K., editors. *The structuring role of submerged macrophytes in lakes*. New York: Springer-Verlag. 1997; 397-406.

Schriver, P.; Bøgestrand, J.; Jeppesen, E.; Søndergaard, M. Impact of submerged macrophytes on fish–zooplankton–phytoplankton interactions: large-scale enclosure experiments in a shallow trophic lake. *Freshwater Biol.* 1995. 33, 255–270.

Shannon, C. E; Weaver, W. The mathematical theory of communication. Illinois: University of Illinois Press. 1963

Sheffer, M.; Carpenter, S.; Foley, J. A.; Folkes C. and B. Walker. Catastrophic shifts in ecosystems. *Nature*. 2001, 413, 591-596.

Søndergaard, M.; Moss, B. Impact of submerged macrophytes on phytoplankton in shallow freshwater lakes. Ecol. Stu. 1998, 131, 115-132.

Steinberg, C. E. W.; Kamara, S.; Prokhotskaya, V. Y.; Manusadzianas, L.; Karasyova, T. A.; Timofeyev, M. A.; Jie, Z.; Paul, A.; Meinelt, T.; Farjalla, V. F.; Matsuo, A. Y. O.; Burnison, B. K.; Menzel, R. Dissolved humic substances – ecological driving forces from the individual to the ecosystem level? *Freshwater Biol.*, 2006, 51, 1189–1210.

Steinberg, C. E. W.; Kamara, S.; Prokhotskaya, V. Y.; Manusadzianas, L.; Karasyova, T. A.; Timofeyev, M. A.; Jie, Z.; Paul, A.; Meinelt, T.; Farjalla, V. F.; Matsuo, A. Y. O.;

Burnison, B. K.; Menzel, R. Dissolved humic substances -ecological driving forces from the individual to the ecosystem level? *Freshwater. Biol.* 2006, 51, 1189-1210.

Steinberg, C. E. W.; Meinelt, T.; Timofeyev, M. A.; Bittner, M.; Menzel, R. Humic Substances, Part 2: Interactions with Organisms. *Environ. Sci. Pollut. R.* 2008, 15, 128-135.

Stepanauskas, R.; Laudon, H.; Jørgensen, N. O. G. High DON bioavailability in boreal streams during a spring flood. *Limnol. Oceanogr.* 2000, 45, 1298-1307.

Stewart A. J.; Wetzel R. G. Influence of dissolved humic materials on carbon assimilation and alkaline phosphatase activity in natural algal-bacterial assemblages. *Freshwater Biol.* 1982, 12, 369- 380.

Strickland, J. D. H.; Parsons, T. R. *A practical handbook of seawater analysis.* Ottawa:Fisheries Research Board Canada Bulletin. 1972; v. 167.

Strome, D. J.; Miller, M. C. Photolytic changes in dissolved organic humic substances. *Int. Ver. Angew. Limnol.* 1978, 20, 1248-1954.

Stumm-Zollinger, E. Effects of inhibition and repression on the utilization of substrates by heterogeneous bacterial communities. *Appl. Microbiol.* 1966, 14, 654-664.

Sun, B. K.; Tanji, Y.; Unno, H. Extinction of cells of cyanobacterium *Anabaena circinalis* in the presence of humic acid under illumination. *Appl. Microbiol. Biotechnol.* 2006, 72, 823–828.

Sun, B. K.; Tanji, Y.; Unno, H. Influences of iron and humic acid on the growth of the cyanobacterium *Anabaena circinalis. Biochem. Eng. J.* 2005, 24: 195-201.

Sun, J.; Liu, D. Geometric models for calculating cell biovolume and surface area for phytoplankton. *J. Plankton Res.* 2003, 25, 1331-1346.

Teixeira-de-Mello F.; Meerhoff, M.; Pekcan-Hekim, Z.; Jeppesen, E. Substantial differences in littoral fish community structure and dynamics in subtropical and temperate shallow lakes. *Freshwater Biol.* 2009, 54, 1202–1215.

They, N. H.; D. M. M. Marques; Jeppesen, E.; Søndergaard, M. Bacterioplankton in the littoral and pelagic zones of shallow lakes. *Hydrobiologia.* 2010, 646, 311-326.

They, N. H.; Marques, D. M.; & Souza, R. S. Lower respiration in the littoral zone of a subtropical shallow lake. *Front. Microbiol.* 2013a, 3, 1-10.

They, N. H.; Marques, D. M.; Souza, R. S.; Rodrigues, L. R. Short-term photochemical and biological unreactivity of macrophyte-derived dissolved organic matter in a subtropical shallow lake. *J. Ecosyst.* 2013b, 316709, 1-9.

Utermöhl, H. Zur Vervolkommung der quantitativen phytoplankton-methodik. *Mitt. Int. Ver. Angew. Limnol.* 1958, 9, 1-38,

Van den Berg, M.; Coops, H.; Simons, J.; Keizer, A. De. Clear water associated with dense *Chara* vegetation in the shallow and turbid Lake Veluwemeer, The Netherlands. In: Jeppesen E., editor. *The structuring role of submerged macrophytes in lakes.* Netherlands: *Springer-Verlag.* 1998; 339-352.

van Donk, E.; Gulati, R. D.; Iedema, A.; Meulemans, J. Macrophyte-related shifts in the nitrogen and phosphorus contents of the different trophic levels in a biomanipulated shallow lake. *Hydrobiologia* 1993, 251, 19–26.

van Donk, E.; Gulati, R. D. Transition of a lake to turbid state 6 years after biomanipulation: mechanisms and pathways. *Water Sci. Technol.* 1995, 32, 197–206.

van Donk, E.; van de Bund, W. J. Impact of submerged macrophytes including charophytes on phyto- and zooplankton communities: allelopathy versus other mechanisms. *Aquat. Bot.* 2002, 72, 261-274.

Van Nes H.; Scheffer M.; Van Den Berg, M.; Copps, H. Aquatic macrophytes: restore, eradicate, or is there a compromise? *Aquat. Bot.*. 2002, 72, 387-403.

Van Nes, E. H.; Scheffer, M.; Van Den Berg, M.; Copps, H. Charisma: a spatial explicit simulation model of submerged macrophytes. *Ecol. Model.* 2003, 159, 103-116.

van Wijk, R. J. Ecological studies on *Potamogeton pectinatus* L. I. General characteristics, biomass production and life cycles under field conditions. *Aquat. Bot.* 1989, 31, 211-158.

Wardle, D. A.; Nilsson, M.-C.; Gallet, C.; Zackrison, O. An ecosystem-level perspective on allelopathy. *Biol. Rev.* 1998, 73, 305-319.

Waridel, P.; Wolfender, J.-L.; Lachavanne, J.-B.; Hostettmann, K. *ent*-Labdane diterpenes from the aquatic plant *Potamogeton pectinatus*. *Phytochemistry*. 2003, 64, 1309-1317.

Weisner, S.; Eriksson, G.; Granéli, W.; Leonardson, L. Influence of macrophytes on nitrate removal in wetlands. *Ambio*. 1994, 23, 363–366.

Wetzel, R. G. Extracellular enzymatic interactions: storage, redistribution, and interspecific communication. In: Chróst, R. J., editor. *Microbial enzymes in aquatic environments*. New York: *Springer-Verlag*. 1991; 6-26.

Wetzel, R. G. Gradient-dominated ecosystems: sources and regulatory functions of dissolved organic matter in freshwaters ecosystems. *Hydrobiologia*. 1992, 229, 181-198.

Wetzel, R. G.; G. E. Likens. Limnological Analyses. New York: *Springer-Verlag*. 2000; 3rd edition.

Willis, R. J. The historical bases of the concept of allelopathy. *J. Hist. Biol.* 1985, 18, 71–102.

Willis, R. J. The history of allelopathy. Netherlands: Springer. 2007.

Wium-Andersen, S.; Christophersen, C.; Houen, G. Allelopathic effects on phytoplankton by substances isolated from aquatic macrophytes (Charales). *Oikos* 1982, 39, 187–190.

Wium-Andersen, S.; Anthoni, U.; Houen, G. Elemental sulphur, a possible allelopathic compound from *Ceratophyllum demersum*. *Phytochemistry*. 1983, 22, 2613.

Wium-Andersen, S. Allelopathy among aquatic plants. *Arch. Hydrobiol. Beih. Ergebn. Limnol.* 1987, 27, 167–172.

Wu, Q. L.; Zwart, G.; Wu, J.; Agterveld, M. P. K.-v.; Liu, S.; Hahn, M. W. Submersed macrophytes play a key role in structuring bacteriplankton community composition in the large, shallow, subtropical Taihu Lake, China. *Environ. Microbiol.* 2007, 9, 2765-2774.

In: New Developments in Allelopathy Research
Editor: Julia E. Price
ISBN: 978-1-63483-390-5

Chapter 6

THE INVASIVE POTENTIAL OF PARTHENIUM WEED: A ROLE FOR ALLELOPATHY

Boyang Shi[1], Zubair Aslam[2] and Steve Adkins[1]
[1]School of Agriculture and Food Sciences,
[2]The University of Queensland, Australian
Department of Agronomy, University of Agriculture, Faisalabad, Pakistan

ABSTRACT

Parthenium weed (*Parthenium hysterophorus* L.) is an herbaceous plant that has now invaded over 44 countries worldwide having an impact upon crop and pasture production, natural community biodiversity and human and animal health. Allelopathy is a well-defined physiological trait that is possessed by some crops and native plant species, but especially by invasive weeds. It is a biological phenomenon involving the release of chemicals that may cause a stimulatory, but more often an inhibitory effect upon another plant's growth, reproduction and survival. Numerous studies have associated the invasive success of parthenium weed with a proficient allelopathic attribute, able to cause an inhibition of either germination, seedling or plant growth, phenological development and/or flowering in a range of agriculturally important plant species, including weeds and some native species.

The presumed allelochemicals are produced by a number of plant parts including the stem, leaves, roots, pollen and other floral parts. The two major allelochemical groups that are thought to be involved are the sesquiterpene lactones and the phenolic acids, and both have been found to be released by living plants and decomposing plant residues as well as being present in rhizosphere soil under a parthenium weed population. Different bioassay approaches have been used to help illucidate the allelopathic nature of this weed. These include laboratory assays using aqueous extracts applied directly to germinating seeds, or those evaluating the impact of parthenium weed seedlings on the growth of neighbouring seedlings, both using a liquid medium (relay seeding bioassay) or in an agar medium (plant box bioassay). Other laboratory bioassays have looked at the impact of parthenium weed litter upon germination and seedling growth (sandwich bioassay) while some glasshouse pot tests carrying sand or soil have also been undertaken looking at the impact of aqueous extracts, leaf litter or rhizosphere soil. Each approach has its own advantages and disadvantages. This chapter will evaluate the evidence and decide if allelopathy makes a major or a minor contribution to the invasiveness of this weed.

1. Parthenium Weed

1.1. Origin and Current Distribution

Parthenium weed (*Parthenium hysterophorus* L.; Asteraceae) is an alien invasive herbaceous weed originating from the tropical and sub-tropical Americas, a region including portions of North, Central and South America, and the Caribbean islands (Dale, 1981). The weed is now found in 87 countries, only 44 of which are considered to be within its native range (Adkins and Shabbir, 2014: Table 1.1).

Parthenium weed is a major weed in Eastern and Southern Africa, Eastern and Southern Asia, Australia and the Pacific (Adkins and Navie, 2006), and it continues to invade, then spread, within other countries that have a suitable climate to support its growth.

This potential new range includes parts of Southeast Asia, West Africa and Eastern Europe (McConnachie et al., 2011). At this point in time, Australia, India and Ethiopia are the most affected countries with much of their cropping and/or grazing lands affected (Tamado et al., 2002a; Lakshmi and Srinivas, 2007).

Parthenium weed can come to dominate in a wide range of different habitats, including grazing and cropping lands, forests, protected areas, along train, river and roadsides, in wastelands and construction areas, in gardens, orchards, and in stockyards etc. (Navie et al., 1996; Navie, 2002). Parthenium weed shows some tolerance to drought and high temperature but is intolerant to heavy shade, and almost all kinds of soils are suitable for its growth, except for saline soils (Lakshmi and Srinivas, 2007). During the very early stages of its growth the leaves produced are egg shaped and hairy with short peduncles (Mahadevappa, 2009).

In the rosette stage, leaves are larger, covered with fine hairs, pale-green in colour, and form a radial pattern close to the soil surface. These leaves can reach up to 10 cm in length and 5 cm in width, sometimes forming a carpet of leaves on the ground. When the plant attains maturity, it can reach a height of 2.5 m, with a deep tap root system and an erect main stem, with many side branches carrying deeply divided leaves (McFadyen, 1992) and numerous flowers producing up to 25,000 seeds per plant (Adkins and Shabbir, 2014).

World-Wide

The weed has now been accidentally introduced into 43 countries around the world including India, Australia, South Africa, Ethiopia, China, Vietnam, several islands in the Pacific and most recently into Malaysia and the United Arab Emirates (Figure 1.1 and Table 1.2). From the major introduction sites, further international and national spread has occurred. The pathways of international spread have become more common place in recent years as travel and international trade activities have increased in frequency. For example, from India, one of the most affected countries, parthenium weed has now spread into most neighbouring countries including Pakistan, Nepal, Bangladesh, and Bhutan, presumably by seed attached to vehicles or as a contaminant of fodder or seed lots (Mishar, 1991; Parker, 1992). Furthermore, in Sri Lanka the weed was probably introduced from India by goats accompanying a military mission in the period 1987-90 (Jayasurya, 2005). In other locations around the globe the weed has been shown to be dispersed nationally by a wide range of biotic and abiotic pathways including as a contaminant in grain and other seed lots, by domesticated, feral and native animals, by vehicles, by wind and by water bodies, including river and flood waters (Adkins and Shabbir, 2014).

Australia

Parthenium weed has been accidentally introduced into Australia on two separate occasions. The first introduction was at a site near Toogoolawah in southeast Queensland, which was reported in 1950, but was likely to have occurred in the mid 1940's (Parsons and Cuthbertson, 2001). The second introduction was at a site near Clermont in central Queensland in 1958 when a contaminated pasture grass seed lot, imported from Texas, USA was sown into a beef pasture (Haseler, 1976).

This second introduction has been the most important one as within 30 years of its arrival it had spread to cover about 10% (*ca.* 17 million ha) of Queensland (Chippendale and Panetta, 1994) and by 2013 it covered more than *ca.* 60 million ha. In Queensland, the weed has been made a class 2 pest plant under the Land Protection Act 2002 and is one of the original 20 Weeds of National Significance (A National Program of the States, Territories and the Australian Government, 2009).

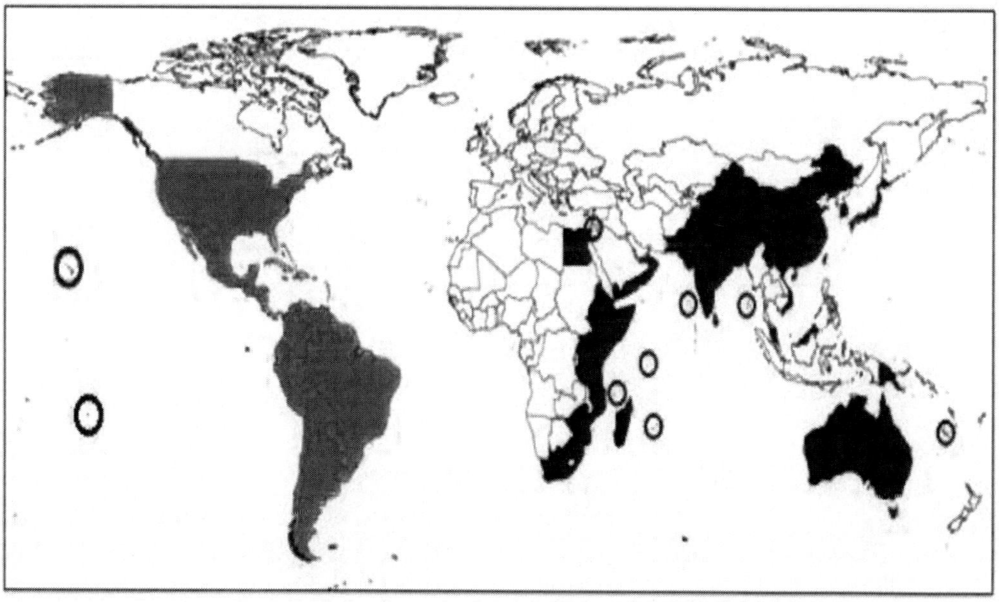

Figure 1.1. The distribution of parthenium weed around the world. The 43 countries shaded or circled in black are considered to be within its introduced range while those 44 shown in grey are within it's a native range (modified from Shabbir, 2012).

Table 1.1. The 44 North and South American countries where parthenium weed is thought to be native, and the reference where this is reported

North America		South America	
Country	Reference	Country	Reference
Anguilla	EPPO, 2014	Argentina	Rollins, 1950
Antigua and Barbuda	USDA-ARS, 2012	Bolivia	Friend, 1983
Aruba	EPPO, 2014	Brazil	Parsons and Cuthbertson,
Bahamas	EPPO, 2014	Chile	2001
Barbados	Dale, 1981	Ecuador	Dale, 1981
Belize	Kbosla and Sobti, 1981	French Guiana	EPPO, 2014
Bermuda	Dale, 1981	Guadeloupe	Dale, 1981
Cayman Island	EPPO, 2014	Guyana	EPPO, 2014
Costa Rica	Parsons and Cuthbertson,	Netherlands	Dale, 1981
Cuba	2001	Antilles	USDA-ARS, 2012
Curacao	Rollins, 1950	Paraguay	Rollins, 1950
Dominica	Dale, 1981	Peru	Dale, 1981
Dominican Republic	Dale, 1981	Surinam	USDA-ARS, 2012
Grenada	Dale, 1981	Turks and Caicos	USDA-ARS, 2012
Guatemala	USDA-ARS, 2012	Islands	
Haiti	Friend, 1983	Uruguay	Rollins, 1950
Honduras	Rollins, 1950	Venezuela	Rollins, 1950
Jamaica	Rollins, 1950		
Martinique	Dale, 1981		
Mexico	Dale, 1981		
Nicaragua	Rollins, 1950		
Panama	Lewis et al., 1988		
Puerto Rico	Hammerton, 1981		
Saint Kitts and Nevis	Friend, 1983		
Saint Lucia	USDA-ARS, 2012		
Saint Vincent and the	EPPO, 2014		
Grenadines	USDA-ARS, 2012		
Trinidad	Friend, 1983		
United States Virgin Islands	USDA-ARS, 2012		
USA	Rollins, 1950		

1.2. Impact

Crop Production

Serious declines in crop yield have been observed in wheat (*Triticum aestivum* L.), sorghum [*Sorghum bicolor* (L.) Moench] and maize (*Zea mays* L.) production (Hassan and Khan, 2010, Tamado et al., 2002a, Towers et al., 1997 and Shabbir and Bajwa, 2006) and in 45 other field crops (Table 1.3). For example, in India, wheat yields declined by as much as 40% (Khosla and Sobti, 1981) and in Pakistan by 46% when parthenium weed was poorly managed (Hassan and Khan, 2010). In Ethiopia, sorghum yield losses were reported to be as high as 60% (Tamado et al., 2002b) while for teff (*Eragrostis tef* Zucc. Trotter) yield reductions have now become a national concern as this crop is a staple food in this country (Tefera, 2002).

Table 1.2. The 43 countries within Africa and the Middle East, Asia and Europe and Oceania where parthenium weed has been introduced, and the reference where this is reported

Africa and the Middle East		Asia		Europe and Oceania	
Country	Reference	Country	Reference	Country	Reference
Comoros	EPPO, 2014	Bangladesh	Navie et al., 1996	Australia	Everist, 1976
Djibouti	Etana et al., 2015	Bhutan	Parker, 1992	Belgium	EPPO, 2014
Egypt	Zahran & Wills, 2009	China	Towers et al., 1977	French Polynesia	Aneja et al., 1991
Eritrea	USDA-ARS, 2012	India	Rao, 1956	Poland	EPPO, 2014
Ethiopia	Medhin, 1992	Israel	Joel & Liston, 1986	Hawaii	Motooka et al., 2003
Kenya	Njoroge, 1991	Japan	Mito & Uesugi, 2004	New Caledonia	MacKee, 1994
Madagascar	Tamado et al., 2002	Malaysia	Karim, 2013	PNG	Anonymous, 2003
Mauritius	EPPO, 2014	Nepal	Mishra, 1991	Tahiti	QLD Government, 2011
Mayotte	USDA-ARS, 2012	Pakistan	Razaq et al., 1994	Vanuatu	Nath, 1981
Mozambique	Tamado & Milberg, 2000	South Korea	Gil Rhee, 2004		
Oman	Kilian et al., 2002	Sri Lanka	Jayasurya, 2005		
Reunion	EPPO, 2014	Taiwan	Peng et al., 1988		
Seychelles	Nath, 1988	Vietnam	Towers et al., 1977		
Somalia	Tamado et al., 2002				
South Africa	Rollins, 1950				
Swaziland	Bowen, 2001				
Tanzania	Clark et al., 2010				
UAE	Tamer et al., 2015				
Uganda	Bisikwa, 2010				
Yemen	Kilian et al., 2002				
Zimbabwe	Tamado et al., 2002				

Table 1.3. The 48 crops that have become infested by parthenium weed, and the reference where this is reported

Crop	Scientific name	Reference
Fibre		
Cotton	*Gossypium hirsutum* L.	Parsons and Cuthbertson, 2001
Grain		
Barley	*Hordeum vulgare* L.	Sriastave et al., 1985
Finger millet	*Eleusine coracana* L.	Bhatt et al., 1994
Maize	*Zea mays* L.	Bhowmik and Doll, 1982
Rice	*Oryza sativa* L.	Singh and Sangeeta, 1991
Sorghum	*Sorghum bicolo* (L.) Moench	Tamado et al., 2002a
Sugarcane	*Saccharum officinarum* L.	Hossain et al., 2015
Wheat	*Triticum aestivum* L.	Singh and Sangeeta, 1991
Vegetable		
Bottle gourd (Calabash)	*Lagenaria siceraria* L.	Hossain et al., 2015
Brinjal (eggplant)	*Solanum melongena* L.	Hossain et al., 2015
Cabbage	*Brassica oleracea* L.	Maharjan et al., 2007
Chili pepper	*Capsicum annuum* L.	Dhawan and Dhawan, 1995
Cucurbit	*Cucurbita sp.*	Hossain et al., 2015
Garlic	*Allium sativum* L.	Hossain et al., 2015
Lettuce	*Lactuca sativa* L.	Wakjira et al., 2009
Okra	*Abelmoschus esculentus* Moench	Dhawan and Dhawan, 1995
Onion	*Allium cepa* L.	Wakjira, 2009
Peas	*Pisum sativum* L.	Sriastave et al., 1985
Potato	*Solanum tuberosum* L.	Hossain et al., 2015
Radish	*Raphanus sativus* L.	Maharjan et al., 2007
Spinach	*Spinacea oleracea* L.	Dhawan and Dhawan, 1995 and Gupta, 1996
Sweet gourd	*Cucurbita maxima L.*	Hossain et al., 2015
Taro (maan kochu)	*Colocasia esculenta* L.	Hossain et al., 2015
Tomato	*Lycopersicon esculentum* Mill.	Mersie and Singh, 1988
Pulse		
Black gram	*Phaseolus mungo* L.	Singh and Sangeeta, 1991
Clover	*Trifolium ambiguum* M. Bieb	Dhawan and Dhawan, 1995
Cowpea	*Vigna unguiculata* L.	Swaminathan et al., 1990
Chickpea	*Cicer arientinum* L.	Singh and Sangeeta, 1991
French bean	*Phaseoulus vulgaris* L.	Madhu et al., 1995
Green gram	*Vigna radiate* L.	Kohli and Rani, 1992
Lentil	*Lens culinaris* L.	Hossain et al., 2015
Mung bean	*Vigna mungo* L.	Kohli and Rani, 1992
Oil seed		
Brassica	*Brassica napus* L.	Singh et al., 2005
Field mustard	*Brassica campestris* L.	Maharjan et al., 2007
Soybean	*Glycine max* L.	Bhowmik and Doll, 1982
Sunflower	*Helianthus annuus* L.	Bhowmik and Doll, 1982
Tree		
Acacia	*Acacia leucophloea* L.	Swaminathan et al., 1990
Forest Gum	*Eucalyptus tereticornis* Sm.	Swaminathan et al., 1990

Crop	Scientific name	Reference
Leucaena	*Leucaena leucocephala* (Lam.) De Wit	Swaminathan et al., 1990
Mango	*Mangifera indica* L.	Hossain et al., 2015
Mulberry	*Morus papyrifera* L.	Khan, 2011
She-oak	*Casuarina equisetifolia* L.	Swaminathan et al., 1990
Other		
Banana	*Musa balbisiana* L.	Hossain et al., 2015
Citrus	*Citrus reticulate* Blanco	Khan, 2011
Coffee	*Coffea arabica* L.	Njoroge, 1986
Tea	*Camellia sinensis* L.	Njoroge, 1986
Tobacco	*Nicotiana rustica* L.	Khan, 2011
Turmeric	*Curcuma longa* L.	Hossain et al., 2015

Pasture and Rangeland Production

Parthenium weed is also a noxious weed in pastures, rangelands and native communities in most countries where it has invaded. For example, in Australia, parthenium weed is estimated to be causing economic losses of more than AUS $100 million annually to the beef industry (Adamson, 1996). Dhileepan (2007) has shown grass species richness to decline as parthenium weed infestation increase in size. Nguyen (2011) found the biodiversity of a pasture community to decline significantly even when parthenium weed density was two plants per m^2. In India the carrying capacity of certain pastures has decreased by as much as 90% in the presence of moderate levels of parthenium weed (Nath, 1981; Shabbir, 2012). While this weed presently reduces the richness and diversity of a grassland community, in the future it is expected to have more of an impact as its growth and seed production have been shown to be increased under a changing climate (Nguyen, 2011).

Human Health

Several diseases can be caused by parthenium weed, and are to be seen in almost all countries where this weed has been introduced. These diseases can be divided into two categories; those that cause impact upon the body (dermatitis) and those that cause impact within the body (McFadyen, 1995). Airborne contact dermatitis is one of the commonest forms of parthenium weed disease and normally appears on the face, particularly on the eyelids and nasolabial folds, upon the neck, arms, legs and feet (Nandhakishore and Pasricha, 1994). This dermatitis can be caused by the direct contact with the weed but also by coming into contact with air-borne particles, including leaf and stem hairs, pollen and other small leaf fragments. In addition, if this plant is touched, within a short period of time white dermatological spots can appear on the hands which can then develop into blisters and ulcers (Warren and Heymann, 2006). The internal form of the disease can be devastating and is caused by air-borne fragments being taken in through the mouth or nasal passages and then into the respiratory tract. Here the fragments can cause allergic rhinitis, bronchitis and/or asthma (Bousquet et al., 1997; Martinez, 2007).

Animal Health

Parthenium weed can have adverse effect upon the health of a wide range of domesticated livestock including cattle, horses, goats, buffalo and sheep. The impacts of the weed can be direct or indirect. The direct effects are often brought about by the consumption of small amount of the weed with non-life threatening outcomes such as diarrhoea, alopecia and certain kinds of dermatitis (Narasimhan et al., 1977). However, these impacts can also be more severe in heavily grazed areas where animals develop life-threatening cases of sever dermatitis, or mouth ulcers and poisoning (Ahmed et al., 1988). The indirect effects that brought on by just a mild level of poisoning include tainting of the meat or milk. These outcomes are thought to be caused in dairy cattle by an alteration of the microbial composition within the rumen which in turn causes a bitter taste to the milk produced (Lakshmi and Srinivas, 2007; Narasimhan et al., 1977; Tudor et al., 1982).

2. Methods Used to Assess Parthenium Weed Allelopathy

2.1. Definition

Allelopathy is a phenomenon observed in many plants that involves the release of chemicals into the local environment by either root exudation, residue or litter decay, leaching or volatilization from aerial parts (Figure 2.1) which can affect the growth and/or development of neighbouring plants. Allelopathy is considered to be an important trait of many invasive weeds but also of some crops and native species. The Novel Weapons Hypothesis suggests that this may be a common approach invasive plant species use to gain an advantage over native plants when invading a new habitat (Ni et al., 2012). The chemicals released are secondary plant metabolites which can be produced from a range of tissues with the plant including the roots, leaves, stems, flowers, seeds and pollen. Such compounds may become altered once in the environment to either gain or lose biological activity (Haig, 2008). The chemical structure of the allelopathic compounds can be categorized into a wide range of types, which include certain organic acids, coumarins and flavonoids; also long-chain fatty acids and polyacetylenes, phenolic acids, quinines, simple unsaturated lactones, steroids and terpenoids, tannins as well as other aliphatic aldehydes, straight-chain alcohols and ketones (Zhao et al., 2010). The production of the major allelopathic substances can then give rise to a range of secondary allelochemicals.

As neighbouring plants can affect each other through either an allelopathic or a resource competition pathway the term interference is often used to describe the combined effects of these two interactions. In most cases when plants are grown together in the glasshouse or in the field, both allelopathy and competition will occur simultaneously and it is extremely difficult to separate these two forms of interaction. Thus, laboratory bioassays are often used as the first step to investigate the presence of allelopathy in a plant-plant interaction study. Replicated laboratory bioassays are thought to avoid the effects of competition, as well as being inexpensive, rapid and easy to run, and can be applied to numerous target species at once. In addition, such bioassays can be used to study both the allelopathic capacity of living plants and their dead tissues.

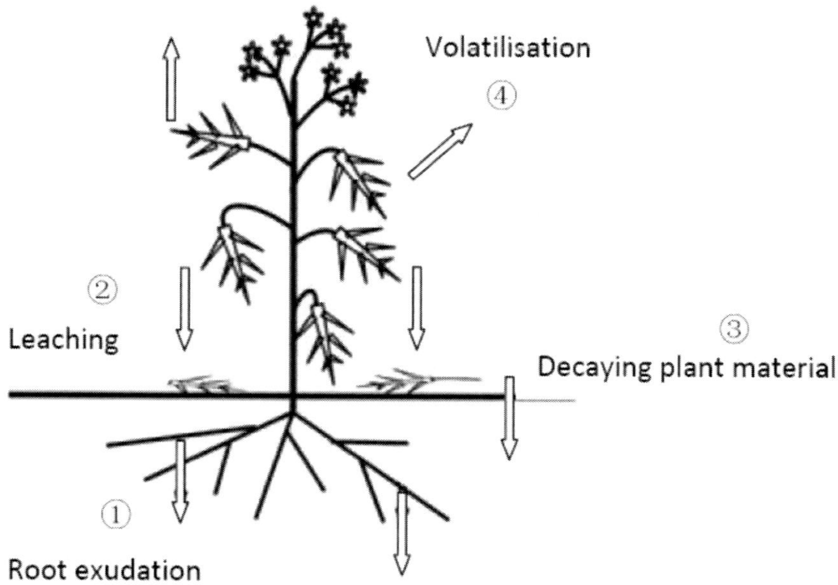

Figure 2.1. Many plants release chemicals into the local environment, either as 1) exudates from roots, or 2) leachates from leaves, stems, flowers, pollen, seed and fruit created by dew and rain, or 3) leachates from decaying plant material also created by rain, or 4) volatiles from leaves. Once in the environment they or their chemical derivatives can produce an allelopathic impact upon the growth of a neighbouring plant (modified from Putnam and Tang, 1986).

2.2. Laboratory

Aqueous Extract Bioassay

Aqueous extract bioassays have been the most commonly used methods to gauge the possible allelopathic activity of parthenium weed. Such bioassays are normally conducted in Petri dishes where the seed of a test species are imbibed in an aqueous extract made from parts of the parthenium weed plant. By using an aqueous extract bioassay, the possible allelopathic activity of parthenium weed has been determined on a large number of agriculturally important species (Table 2.1) with the common result of germination and/or seedling growth being inhibited (e.g., Singh, 1991; Gupta, 2000; Batish et al., 2002a; Channappagoudar et al., 2005; Wakjira et al., 2005; Maharjan et al., 2007; Bhawana et al., 2009; Demissie et al., 2013; Hu et al., 2013 and Babu et al., 2014) but few studies have been undertaken on pasture or native plants. The advantages of this kind of bioassay is that it is rapid and simple to undertake and can be used as a preliminarily test to determine the possibility of parthenium weed being allelopathic against a certain test plant before further investigation is undertaken. The disadvantages of the aqueous extract bioassay is that a) during the preparation of the aqueous extract certain enzymes, salts, amino acids, and nitrogen compounds can be released that then cause a false positive result (Chou and Muller, 1972; Inderjit and Dakshini, 1995) and b) much higher doses than is normally produced under natural conditions are often applied to the seeds, again causing a false positive result. It is therefore essential that an extract bioassay be followed by further approaches in the glasshouse or under field conditions to confirm the presence of an allelopathic effect.

Table 2.1. A selection of studies where aqueous extracts were taken from living parthenium weed plant parts and applied to the seed of test species in a germination laboratory bioassay test, and the outcomes of those tests

Source of parthenium weed	Plant parts used and drying conditions	Assay approach	Receptor species	Incubation condition	Duration of study	Salient results	Reference
Unknown	Whole plant, Air dried then separated into root and shoot and dried in an oven (70°C) for 8 hours, ground and sieved through 2 mm mesh.	10 g of ground material was placed into 100 mL distilled water for 24 h at 23°C, shaken for 20 minutes then filtered.	Seeds of rice, black gram, wheat and chickpea were sown into 10 cm Petri dishes and imbibed with 1, 5 or 10% of the filtered extract.	23±2°C	6 days	Shoot and root extracts reduced the germination of all species. The 5% aqueous extract affected legumes more than cereals. Both extracts reduced shoot and root growth of the crop seedlings.	Singh and Sangeeta, 1991
Roots were collected from plants of different ages growing in the field	The roots from juvenile, rosette, vegetative and mature plants and then oven dried (80°C) for 96 hours and then powdered	250g powdered roots were soaked in 3L of distilled water for 48 hours then filtered	50 soaked seeds of cluster bean, okra and spinach were placed into 9cm Petri-dishes and imbibed with 1, 4 or 8% of the filtered extract.	Laboratory conditions	7 days	The root extracts inhibited the germination and dry accumulation of all species. Vegetative stage extracts caused minimum inhibition while rosette stage caused maximum inhibition with okra.	Gupta, 2000
Leaves collected from the field.	Shade-dried and powdered and then parthenin was extracted.	The extracted parthenin was made into a range of concentrations and used in a germination (38-1,900 µM), growth (190-760 µM) and soil experiment (190-760 µM).	Wild oat and cobbles pegs 25/50, 15/30 seeds used in the germination and growth/soil experiments, respectively.	25±2°C, 16/8 h light/dark photoperiod (150 µmol $m^{-2} s^{-1}$), relative humidity 75%.	7 days for germination, 10 days for growth and 14 days for the soil experiment.	Germination was decreased with the increasing concentration of parthenin. Root growth was inhibited more than shoot growth in the pot test. Chlorophyll content was reduced by all parthenin treatments.	Batish et al., 2002a

Source of parthenium weed	Plant parts used and drying conditions	Assay approach	Receptor species	Incubation condition	Duration of study	Salient results	Reference
Unknown	Leaf samples were collected prior to flowering and dried.	Samples were ground and soaked overnight in 100 mL of distilled water. The concentration of aqueous leaf extracts were made at 5 and 10%.	Sorghum, wheat, green gram, soybean, sunflower and peanut.	Laboratory conditions	9 days	The 10% extract reduced the germination compared to the 5% extract. Soybean was the most affected by the extracts.	Channapp-agoudar et al., 2005
Plants were harvested from the field at the flowering stage.	Root, stem, leaf and flower parts were dried at 70°C for 24 h then ground in a mill	0, 5, 10 or 15 g of ground materials were shaken in 100 mL of distilled water for 24 h at room temperature (c. 22°C)..	Lettuce seeds were placed on filter paper with 10 mL of the four concentrations of extract.	Room temperature (21-22°C) with constant fluorescent light	7 days	All three flower extracts, the 10 and 15 g leaf extracts and the 15 g stem extract inhibited germination of lettuce	Wakjira et al., 2005
Fresh leaves were collected at the vegetative stage of growth from the roadside.	Leaf tissue was air dried in the shade for 7 days then stored in a plastic bag for a month at room temperature (25°C) then ground.	10 g of ground leaves were shaken with 100 mL of distilled water for 24 hours in the dark (25°C). Aqueous extracts of 2, 4, 6, 8 and 10% were prepared.	10 seeds of rice, maize, wheat, radish, mustard, cauliflower, wormwood and Crofton weed placed into Petri dishes	An ambient temperature (25°C mid-day) with natural light	7 days	Germination, root and shoot elongation and biomass were inhibited by the 10% aqueous extract except for maize.	Maharjan et al., 2007
Plants were collected at the time of seeding from a crop field.	Whole plants	10 g of fresh plant material was added to 100 mL of sterile distilled water and soaked for 24 h at room temperature.	Seed of different wheat varieties were placed into Petri dishes with different concentrations of the weed extract.	Laboratory conditions	8 days	All extract concentrations reduced germination and radicle elongation of those seeds that did germinate. Dry weights were also reduced with all extracts.	Bhawana et al., 2009

Table 2.1. (Continued)

Source of parthenium weed	Plant parts used and drying conditions	Assay approach	Receptor species	Incubation condition	Duration of study	Salient results	Reference
Unknown	Root, shoot and leaf tissues were dried in an oven (60°C) till a cons-tant dry weight was obtained, then ground	Either 1, 2, 4, 8, or 16 g of powder from each plant part were added to 400 mL of distilled water.	Onion and bean seed	Laboratory conditions	4 days	The germination percentage and seedling growth of onion were decreased significantly with the increasing extracts concentration.	Demissie et al., 2013
Fresh leaves were collected at the flower stage from the field.	Fresh leaves were cut into 1-2 cm pieces.	The leaf pieces were soaked in distilled water in the ratio of 1:1 (w/v). From this extracts of 5, 25, 50 and 75% were prepared.	25 seeds of plantain and hawkweed were placed in 9 cm glass Petri dishes.	25/20°C day/ night thermos-period with a 12/12 a photo-period with 70% relative humidity	7 days	The 25% extract inhibited germination while the 5% extract had no effect. Shoot lengths of the species decreased with increasing extract concentration.	Hu et al., 2013
Leaves were collected at the start of the flowering stage from a crop field.	Leaves were chopped up, then dried (75°C) for 48 hours and then ground to a fine powder.	The powdered material was mixed with 100 mL distilled water and made into concentrations of 0.5, 1.0 and 5.0%.	10 seeds of black gram were placed into Petri dishes.	22°C with 70% relative humidity, in dark conditions	28 days	All the extracts inhibited the germination of black gram, which was reduced with the increased concentrations.	Babu et al., 2014

Furthermore, it is very important to include at least one aqueous extract from a non allelopathic plant so it can be clearly demonstrated that the effect is due to the allelopathic substance(s) extracted and not to do with the myriad of other primary plant products that are also released during the extraction process.

Relay Seeding Bioassay

A more biologically relevant bioassay called the "relay seeding technique" has been developed by Navarez and Olofsdotter (1996) for the evaluation of natural released allelopathic substances from plant roots. In this bioassay the seed of the test species are sown next to those of pre-germinated parthenium weed seedlings and the two grown together for 10 days in a well-watered Petri dish enclosed in a transparent growth box. Over the young seedlings is placed a washed granular Perlite to provide a physical support for their further growth and to reduce evaporative water loss from the Petri dishes. By using the relay seeding bioassay, the possible allelopathic activity of parthenium weed has been demonstrated (Belgeri, 2012), with a 15% reduction in the root length of curly windmill grass (*Enteropogon acicularis* L.) and buffel grass (*Cenchrus ciliaris* L.; Table 2.2). However, it was interesting to note that the root length of Rhodes grass (*Chloris gayana* L.) was increased. The advantages of the relay seeding bioassay is that it is more biologically relevant than the aqueous extract bioassay, rapid, simple and the test successfully eliminates competition for water, nutrients, and light (Navarez and Olofsdotter, 1996). One drawback of the bioassay is that any dead or dying test plant seedlings may release chemicals that could interfere with the bioassay.

Plant Box Bioassay

Another biologically relevant bioassay called the "plant box method" has been developed by Fujii (1992) and Fujii et al. (1992) for the evaluation of the natural release of allelopathic substances from plant roots. In this bioassay parthenium weed is grown in sand culture for 2 to 4 weeks with appropriate application of irrigation water and nutrients before seedlings are carefully removed, washed with distilled water, blotted dry and inserted into a nylon mesh tube placed in one corner of a Magenta box. The box and tube are then filled with agar and the seed of the test species placed onto the agar surface at different distances from the parthenium weed seedling. The inhibitory effect of parthenium weed seedlings can then be determined by measuring the impact on root and shoot growth of the test species (Table 2.2). By using the plant box method, the possible allelopathic activity of parthenium weed has been demonstrated by Belgeri (2013), Shi (2012) and Shinwari et al. (2013a; b). These studies have reported a greater inhibition of seedling growth when the test species seedlings (lettuce, curly windmill grass and cotton panic grass) are sown close to the parthenium weed seedling with a decrease in effect as the distance increases. By using this test it has been shown that parthenium weed shows only a moderate allelopathic effect against lettuce, which is significantly less than that achieved by other well established allelopathic plants (sweet clover and four o'clock plant; Shinwari et al., 2013b). The advantages of this bioassay are that it can be undertaken easily in the laboratory and it is also an efficient technique for testing many species rapidly. The disadvantages are that the parthenium weed seedlings need to be of equal size before being transferred into the plant box and this can be difficult to achieve. In addition, big seeded species cannot be used as test species as there is not enough room in the plant box to sow such seeds. Furthermore, it is desirable to include at least one known allelopathic and one non allelopathic plant in each test so the level of the allelopathic effect of the test plant can be determined against known standards.

Sandwich Bioassay

Another biologically relevant bioassay called the "sandwich method" has been developed by Fujii (2003) for evaluating the natural release of allelopathic substances from plant residues. In this bioassay plant leaf material is added into half of the wells of a six-well multi-dish plastic plate and sandwiched between two layers of agar. On to the top of the agar is then sown a number of seeds of the test species. After 3 days, the radicle and hypocotyl lengths are measured and the seed germination percentage are recorded (Table 2.3). Shinwari et al. (2013b) showed that parthenium weed litter had a relatively high allelopathic effect on lettuce seedling growth (86% growth reduction) as compared with other more allelopathic species (98% achieved by sweet clover and 87% by four o'clock plant). Similar results have also been obtained by Shi (2012) who found a range of pasture grasses to be inhibited in their growth by dried parthenium weed leaf litter. In this study 50 mg of leaf litter gave a greater inhibition in growth than 10 mg of leaf litter, and the native species (curly windmill grass and cotton panic grass) were inhibited more than the introduced species (buffel grass and Rhodes grass). The advantages of this bioassay are that it is rapid and easy to undertake and can be applied to numerous target species at once. The disadvantages are that the seeds of the test species need to be of a suitable size (cannot be too big or too small), and have high viability and no dormancy. In addition, the parthenium weed litter needs to be collected fresh just before the experiment, which may be difficult at certain times of the year. Furthermore, it is desirable to include at least one known allelopathic and one non allelopathic plant in each test so the level of the allelopathic effect of the test plant can be determined against known standards.

2.3. Glasshouse Screening Bioassay

There are three basic kinds of bioassay that utilize potted plants; those that incorporate litter from the potentially allelopathic plants into the soil substrate in which the test plants are to be sown, those that apply aqueous extracts to the test plants already growing in soil and those that grow plants in soil that has been taken from around the roots of living field-growing parthenium weed plants.

In the first approach, different amounts of parthenium weed leaf biomass (1, 2 or 5 g dry weight kg^{-1} soil) were incorporated in to a soil into which a number of wheat and pea cultivars were sown. In most cases the growth of the cultivar was inhibited by *c.* 25% (Gupta and Narayan, 2010).

In the second kind of pot bioassay, a 10% aqueous leaf extract made from parthenium weed was shown to cause the root growth inhibition of wheat with a 20% strength aqueous extract causing an even greater growth inhibition of roots as well as that of shoots (Dongre and Singh, 2007).

In the third kind of pot bioassay (Safdar et al. 2014) the rhizosphere soil collected from different parthenium weed infested areas was tested to see if it could support the germination and growth of maize (Table 2.4).

Table 2.2. A selection of laboratory bioassay tests where living parthenium weed seedlings are grown alongside test species either growing in a Petri dish (relay seeding test) or in agar (plant box test)

Source of parthenium weed	Plant parts used and drying conditions	Assay approach and Receptor species	Incubation condition	Duration of study	Salient results	Reference
Seed collected from field populations	Seeds were germinated in a Petri dishes lined with a layer of moistened Geowool.	Relay seeding test: 10 seeds of buffel grass, Rhodes grass, desert blue grass, curly windmill grass, cotton panic grass, weeping grass, dwarf amaranth, mikania vine and Cinderella weed were individually sown next to 10 relative uniform 10 day old parthenium weed seedlings.	25/20°C day/night temperature with a 12/12 h photoperiod	27 days	Shoot length was not significantly affected. Root length of curly windmill grass, buffel grass and dwarf amaranth were all reduced and Rhodes grass, weeping grass and Cinderella weed was increased, respectively.	Berlgeri et al., 2012
Seed collected from field populations	Seedlings of parthenium weed were grown in sandy soil contained in plastic pots placed in a glasshouse for 10 or 20 days in different studies	Plant box test: The seedlings were then positioned into a series of root-zone separating tubes which were placed into Magenta boxes. The boxes and root-zone separating tubes were then partly filled with an agar solution. Curly windmill grass, cotton panic grass, weeping grass, creeping blue grass, Rhodes grass, and buffel grass seeds were sown onto the agar surface at different distances from the test plants.	25 ± 2°C day/night temperature with a 12/12 h photoperiod (ca. 100 µmol m^{-2} s^{-1}), relative humidity of 70/75% day/night.	7 days	The linear regression analysis showed a good correlation between shoot and root growth with distance from the parthenium weed seedling for the majority of the native species. 20 days old seedlings showed more inhibition than the 10 days old seedlings, and native species were inhibited more than the introduced species.	Shi, 2012
Seeds of 38 weed species including parthenium weed were field collected	Seedlings were grown in a sandy soil contained in plastic pots placed in a glasshouse for either 14, 21 and 28 days (depending on species).	Plant box test: Seedlings of all species were placed into root zone separating tubes and placed into Magenta boxes which were partly filled with liquid agar. Lettuce seeds were sown onto the agar surface at different distances from the 38 test plants.	25/20°C day/night temperature with 12/12 h photoperiod (200µmol m^{-2} s^{-1})	5 days	Parthenium weed showed a moderate allelopathic capacity compared to all 38 species. Sweet clover and four o'clock plant showed the most inhibition on the growth of lettuce seedlings.	Shinwari et al., 2013b

Table 2.2. (Continued)

Source of parthenium weed	Plant parts used and drying conditions	Assay approach and Receptor species	Incubation condition	Duration of study	Salient results	Reference
Seed collected from field populations	Seedlings of parthenium weed were grown in sandy soil contained in plastic pots placed in a glasshouse for 10 days	Plant box test: The seedlings were then placed into a series of root-zone separating tubes and placed into Magenta boxes. The boxes and root-zone separating tubes were then partly filled with an agar solution. Curly windmill grass, cotton panic grass, weeping grass, forest bluegrass, creeping blue grass, Rhodes grass, buffel grass, green panic grass, tall finger grass, siratro and butterfly pea seeds were sown onto the agar surface at different distances from the test plants.	25 ± 2°C day/night temperature with a 12/12 h photoperiod (ca. 100 µmol m^{-2} s^{-1}), relative humidity 70/75% day/night.	5 days	The linear regression analysis showed a good correlation between shoot and root growth with distance from the parthenium weed seedling for the majority of the native species: curly windmill grass, cotton panic grass and weeping grass. But less so for the introduced species	Belgeri, 2013

Table 2.3. A selection of laboratory bioassay tests where parthenium weed residues are dried and added to agar to form a sandwich and onto which are sown the seed of test species

Source of parthenium weed	Plant parts used and drying conditions	Assay approach	Receptor species	Incubation condition	Duration of study	Salient results	Reference
Large lower leaves collected from vegetative plants growing in a growth chamber.	Leaves were dried (35°C) for 4 days then stored at 15°C and 15% relative humidity.	Autoclaved agar (0.75%) was added to the six wells of a series of multi-dish plastic plates. The dried and broken leaf material (50 mg) was then added to the agar surface in three wells. On to these and all other wells was added a further 5 mL of agar to form a sandwich.	Five seeds of curly windmill grass, cotton panic grass, weeping grass, creeping blue grass, buffel grass and Rhodes were placed onto the agar surface of each well.	25±1°C with 12/12 h photoperiod	7 days	The shoot and root length of all test species were significantly inhibited especially for the native species	Shi, 2012
Shoots collected from in vitro grown seedlings.	Seedlings of parthenium weed, yellow sweet clover, honey clover, thorn apple and four o'clock plant were dried at 60-70°C for 24 h	Agar was added to all six wells of a multi-dish plastic plate. Broken leaf material (10 or 50 mg) from all species was then added to the agar surface of three wells. On to these and all other wells was added a further 5 mL of agar to form a sandwich.	Lettuce	25°C in the dark.	3 days	Sweet clover was the most inhibitory upon the radicle growth of lettuce followed by honey clover, thorn apple and four o'clock plant. Parthenium weed was moderate in its allopathic capacity compared to these and 38 other species.	Shinwari et al., 2013b
Leaves collected from field.	Fresh leaves of parthenium weed, cocklebur and marijuana were dried at 60°C.	Agar was added to all six wells of a multi-dish plastic plate. Broken leaf material (5, 10 or 50 mg) was then added on to the agar surface of three wells. On to these and all other wells was added a further 5 mL of agar to form a sandwich.	Lettuce	25°C in the dark.	3 days	The inhibition of lettuce seedling growth was increased with the increased concentration of parthenium weed leaf litter. Cocklebur and marijuana showed more inhibition on the lettuce growth than parthenium weed.	Akhtar et al., 2014

Table 2.4. A selection of pot culture bioassay used for the allelopathic study of parthenium weed describing the donor tissues, the test species and outcomes

Source of parthenium weed	Plant parts used and drying conditions	Assay approach	Receptor species	Incubation condition	Duration of study	Salient results	Reference
Residues collected from the field	The residues were shade-dried, then powdered.	Litter incorporation: 0.5, 1.0, 2.0 or 3.0 g of both unburnt and burnt plant residue was mixed into 100 g of soil and moistened with 100 mL of distilled water.	20 radish and 10 chickpea seeds were sown into the soil containing the residues	Ambient temperature (25°C) in an incubator with a 16/8 h light/dark photoperiod and a RH of 75%	7 days	The extracts prepared from both unburnt and burnt residues decreased the growth of both radish and chickpea, but the burnt extracts were more phytotoxic than the unburnt, and the amount of phenolic acids produced was significantly higher.	Singh et al., 2003
Plants collected from the field	Unknown	Litter incorporation: Parthenium weed shoot material was mixed with materials from three common weeds and banana in the ratio of 1:0, 1:1, 1:2, 1:3, 1:4 and 1:5 (w/w) and composted	10 lettuce seeds were sown into the compost produced from the mixtures and all ratios	Glasshouse conditions	7 days	Composting significantly decreased the inhibitory effects of parthenium shoot material on lettuce seed emergence and growth. Composting with other plant material further reduced the impact on emergence and growth	Wakjira et al., 2009
Fresh leaves collected from the field	Leaves, oven dried at 60°C for 48 h	Litter added to soil: 0, 1, 2, 5 g of powdered leaf material was added to each kg of soil collected from a crop field and contained within pots.	10 seeds of wheat and 6 seeds of pea were sown separately in the amended soils. Shoot length and basal diameter, number of leaves produced measured over time and shoot biomass at harvest..	Ambient temperature (25 -12°C) in a shade-house with natural light and a maximum and minimum RH of 72 and 45%	20, to 110 days for wheat, and 27 to 83 days for pea.	No significant effect on germination of either species. Treated plants showed higher initial growth however the number of leaves produced did not increase by the time of harvest.	Gupta and Narayan, 2010

Source of parthenium weed	Plant parts used and drying conditions	Assay approach	Receptor species	Incubation condition	Duration of study	Salient results	Reference
Unknown	Leaf extracts	Aqueous leaf extracts: Pots were filled with 2 kg of garden soil. Leaf extracts (10 and 20%) were applied at the rate of 200 mL pot^{-1} at 7 days intervals	Wheat	Unknown	90 days	The parthenium weed leaf extracts caused the greatest inhibition on shoot biomass. The growth of wheat was significantly reduced with all leaf extracts.	Dongre and Singh, 2007
Rhizosphere soil collected from various field locations	Soil dried at room temperature and sieved before used	Rhizospheric soil: 11 cm diameter and 5 cm depth plastic pots were filled with 350 g of soil from each location. Soil from weed free sites was used as a control.	10 seeds of maize were sown into the soil	29-37°C with a photoperiod of 13 hours	7 days	Maximum reductions in germination and seedling growth were found with rhizosphere soil collected from plants growing near a water channel, while less inhibition was noted in rhizosphere samples taken from a dryer field location.	Safder et al., 2014

Table 2.5. A selection of field assays used for the allelopathic study of parthenium weed describing the condition of the plants, the test species and the outcomes

Source of parthenium weed	Plant parts used and drying conditions	Assay approach	Receptor species	Incubation condition	Duration of study	Salient results	Reference
Unknown	Field growing plants	Field bean seeds were sown at 10, 20 and 30 cm distance apart in six radial rows spanning out from healthy parthenium weed plants which had just started flowering.	Bean	Field conditions of 18-29°C and 63% relative humidity	30 days	Parthenium weed caused retarded growth and nodulation of the bean plants, the inhibition decreased with distance from the parthenium weed plant	Kanchan and Jayachandra, 1979
Unknown	Field growing plants	Densities of 0, 3, 7, 13, 27, 53 and 100 parthenium weed plants m^{-2} were created by transplanting seedlings into a sorghum field at the time of sowing	Sorghum	Two fields, one with a mean annual temperature of 20°C, the other 26°C	21, 42 and 63 days	69% of sorghum yield loss could be created by very low weed densities (three plant m^{-2}).	Tamado et al., 2002a

The results demonstrated that both germination and seedling growth was reduced with maximum reductions occurring with a rhizosphere soil collected from plants growing near a water channel, while lower inhibition was seen with a rhizosphere soil taken from a dryer location, indicating that plant health may play an important role in the strength of the allelopathic effect. The advantages of pot bioassays are that they are more environmentally relevant than laboratory tests, and a relatively large number of test plants can be preliminary screened in a small glasshouse space, before field studies are undertaken.

The disadvantages of the rhizosphere bioassay are that the soils needs to be collected carefully from close to the parthenium weed plants and it can be very time consuming to measure all roots and hypocotyls lengths and to take dry biomass weight of the test species when growing in soil.

2.4. Field Screening Method

Field screening assays are rarely used due to the fact that it is very difficult to separate out an allelopatic response from one due to resource competition. By using this approach the inhibition of bean plant growth was shown to decrease as the distance from the parthenium weed plants increased (Kanchan and Jayachandra, 1979).

Another study used different densities of parthenium weed (3 to 100 plant m^{-2}) deliberately sown into a sorghum field (Tamado et al., 2002) to increase yield loss to 69%, even in the plots where very low densities were created (three plants m^{-2}).

The advantage of field screening is that crop yield data can be generated, which helps in an understanding of the field situation. The drawbacks of this technique are that it's difficult to separate the effects of allelopathy from competition, and the unpredictable environmental conditions under which these tests are often undertaken, including extremes of temperature, soil moisture, soil microorganism, nitrogen concentration also interfere with the collection of data.

3. The Nature of Parthenium Weed Allelopathy

3.1. Source and Nature of the Allelopathic Chemicals

Parthenium weed is thought to produce allelochemicals from a range of its tissues including roots, leaves, stems, flowers, fruit and pollen (Table 3.4). One major source of these chemicals is undoubtedly the aerial parts where certain allelochemicals are contained within the multitude of trichrome hairs that cover the surface of leaves and stems, and can be released directly to the soil during rainfall events (Yaduraju et al., 2005). In addition, stems, leaves and flowers (Ulrichs et al., 2010) of parthenium weed have been shown to contain water-soluble sesquiterpenes lactones (*viz.* parthenin, coronopilin) as well as phenolic acids (*viz.* ferulic acid, vanicillic acid, anicic acid and fumaric acid) which have been shown to have a negative effect upon the germination and growth of a range of test species (Table 3.1, 3.3). On the other hand, parthenium weed roots are known to contain several phenolic compounds (Table 3.4), but not parthenin or coronopilin.

The chemical content of leaf litter is not well established but it is thought to contain both parthenin and coronopilin (Kohli et al., 2006; Safde et al., 2014) and possibly phenolic acids, and therefore could ensure that growth inhibition of neighbouring plants continues all year long. There may be two different mechanisms by which parthenium weed leaf residues reduce plant growth (Wondimagegnehu and Megh, 1987).

Firstly, by the direct inhibition of cell division which can be achieved by phenolic acids, or by the inhibition of root cell elongation, which can be achieved by a wide range of parthenium weed allelochemicals (Bhowmik and Doll, 1982).

3.2. Kinds of Plants Affected

Crops

A wide range of crop species have been shown to have their germination and/or growth reduced by aqueous plant extracts (Singh, 1991; Gupta, 2000; Maharjan et al., 2007; Hu et al., 2013; Babu et al., 2014) and occasionally a reduction in flower and seed production has been observed (Singh et al., 2005; Tefera, 2002; Table 3.1).

All kinds of crop life forms seem to be affected, including short lived annuals to long lived woody perennials.

In addition, crops from a diverse range of families have been reported to be affected by parthenium weed and not just close relatives, or other dicotyledonous plants. Furthermore, crops can also be affected under different environmental conditions, either in the warm or cool season of the year. The growth of some important crops, such as wheat, sorghum, maize and rice, have been shown to be inhibited by parthenium weed both in pot and field experiments; however, it is very hard to say that this is due to allelopathy alone.

Pastures

Only a few pasture species have been examined and shown to have their growth and/or development reduced by parthenium weed (Table 3.2). Of the plants that have been tested most come from the Poaceae family. Some are introduced grasses but it is the native species that seem to be most sensitive.

The growth of some important pasture plants, such as ryegrass, Rhodes grass and buffel grass, have been shown to be inhibited by parthenium weed in laboratory tests.

Other Weeds

Two important environmental weeds and 11 cropping weeds have all been shown to be negatively affected by aqueous extracts made from parthenium weed (Table 3.3). All kinds of weed life form seem to be affected, including short lived annuals to long lived perennials. In addition, weeds from a diverse range of families have been reported to be affected by parthenium weed and not just close Asteraceae relatives. The growth of some important environmental weeds such as Crofton weed and mikania vine, and agricultural weeds such as wild oat and goose foot, have been shown to be inhibited by parthenium weed.

Table 3.1. Crop species that have been reported to be affected by parthenium weed's allelopathic capacity

Crop	Scientific name	Produced by	How tested	Response	Reference
Grain					
Pearl millet	*Pennisetum glaucum* L.	Leaf extract	Petri dish	Radical and shoot growth reduced	Sajjan and Pawar, 2005
Sorghum	*Sorghum bicolo* (L.) Moench				Channappagoudar et al., 2005
Oat	*Avena sativa* L.			Growth and dry weight reduced	Pandey et al., 2011
Rice	*Oryza sativa* L.	Aerial part extract		Germination and shoot growth reduced	Singh and Sangeeta, 1991
Wheat	*Triticum aestivum* L.				
Finger millet	*Eleusine corccana* L.			Germination reduced	Bhatt et al., 1994
Barley	*Hordeum vulgare* L.	Root extract		Germination and shoot growth reduced	Rashid et al., 2008
Maize	*Zea mays* L.				
Vegetable					
Cabbage	*Brassica oleracea* L.	Leaf extract	Petri dish	Germination and shoot growth reduced	Maharjan et al., 2007
Radish	*Raphanus sativus* L.				Maharjan et al., 2007
Spinach	*Spinacea oleracea* L.			Seedling growth reduced	Dhawan and Dhawan, 1995
Groundnut	*Arachis villosulicarpa* Hoehn	Leaf extract	Petri dish	Germination and shoot growth reduced	Channappagoudar et al., 2005
Lettuce	*Lactuca sativa* L.	Leaf litter	Sandwich	Shoot and root length inhibited	Shinwari et al., 2013
Onion	*Allium cepa* L.	Aerial part extract	Petri dish	Germination and growth reduced	Demissie et al., 2013
Tomato	*Lycopersicon esculentum* Mill.	Soil residue	Pot	Interfere the nutrient content	Mersie and Singh, 1988
Black gram	*Phaseolus mungo* L.	Aerial part extract	Petri dish	Germination and shoot growth reduced	Singh and Sangeeta, 1991
Chick pea	*Cicer arientinum* L.				
French bean	*Phaseoulus vulgaris* L.			Germination and growth reduced	Demissie et al., 2013
Cowpea	*Vigna unguiculata* L.	Leaf extract			Swaminathan et al., 1990
Green gram	*Vigna radiate* L.			Germination and shoot growth reduced	Channappagoudar et al., 2005
Mung bean	*Vigna mungo* L.			Germination and growth reduced	Parthasarathi et al., 2012
Peas	*Pisum sativum* L.			Seedling growth reduced	Sriastave et al., 1985

Crop	***Scientific name***	**Produced by**	**How tested**	**Response**	**Reference**
Pigeon pea	*Cajanus cajan* L.	Leaf, stem, root extract			Satsangi et al., 2002
Oil seed					
Brassica	*Brassica napus* L.	Aqueous extract	Petri dish	Germination and seedlings growth reduced	Singh et al., 2005
Mustard	*Brassica campestris* L.	Leaf extract		Germination and shoot growth reduced	Maharjan et al., 2007
Soybean	*Glycine max* L.				Channappagoudar et al., 2005
Sunflower	*Helianthus annuus* L.				
Peanut	*Arachis hypogea* L.		Petri dish	Germination and growth reduced	Parthasarathi et al., 2012
Pumpkin	*Cucurbita moschate* L.	Root exudates	Greenhouse	Seedlings growth reduced	Guzman, 1988
Tree					
Acacia	*Acacia leucophloea* L.	Leaf extract	Petri dish	Germination and growth reduced	Swaminathan et al., 1990
She-oak	*Casuarina equisetifolia* L.				
Forest Gum	*Eucalyptus tereticornis* Sm.				
Leucaena	*Leucaena leucocephala* (Lam.) De Wit				

Table 3.2. Pasture species that have been reported to be affected by parthenium weed's allelopathic capacity

Species	Scientific name	Produced by	How tested	Effect	Reference
Ryegrass	*Lolium multiflorum* Lam.	Shoot extract and shoot residue	Petri-dish sand culture	Root growth reduced	Mersie and Singh, 1987
Summer grass	*Digitaria sanguinalis* L.	Leaf and flower extracts	Petri dish	Seedling growth reduced	Chen et al., 2011
Buffel grass	*Cenchrus ciliaris* L.	Living seedlings	Relay seeding	Shoot growth of desert bluegrass, cotton panic grass, and curly windmill grass was reduced and root growth of curly windmill grass, and buffel grass was reduced	Belgeri et al., 2012
Cotton panic grass	*Digitaria brownii* Hughes				
Curly windmill grass	*Enteropogon acicularis* L.				
Desert bluegrass	*Bothriochloa ewartiana* L.				
Rhodes grass	*Chloris gayana* Kunth.				
Weeping grass	*Microlaena stipoides* R.Br	Living seedlings	Plant box	Shoot and root growth reduced	Shi, 2012
Creeping blue grass	*Bothriochloa insculpa* L.				
Legume	*Alysicurpus glumaceae*	Seed and leaf extracts	Petri dish	Germination, shoot and root length, fresh and dry weight inhibited	Msafiri et al., 2013

3.3. Which Allelochemicals Are Involved?

It has been suggested on a number of occasions (e.g., Batish et al., 2012) that the allelopathic characteristic of parthenium weed is, in part caused by the presence of the sesquiterpene lactone parthenin, which is commonly found in the trichrome hairs that are present on many aerial parts of the weed, including the stem, leaves, and flowers, in leaf litter and in pollen grains (Batish et al., 2002b). Parthenium weed is also known to release several water soluble phenolic acids (including caffeic, vanillic, coumaric, anisic, feluric and chlorogenic acid) from its fruit, leaves, and especially from its roots (Kanchan and Jayachandra, 1980; Mersie and Singh, 1988; Picman and Picman, 1984) all of which might also be involved in the allelopathic response.

Evidence of the Role of Parthenin in Living Plants

The germination and radicle growth of many plant species had been shown to be inhibited by parthenin (Guzman, 1988). Reinhardt et al. (2006) have also shown the parthenin content in leaves increases with age, and peaks at early flowering. In Ethiopia, Ulrichs et al. (2010) indicated that the parthenin content was highest in flowers and leaves, and in all habitats sampled. Interestingly though, a study by Beltz et al. (2001) has shown that parthenin is rapidly degraded in certain kinds of soil with 50% of it being lost after just 59 hours under typical field conditions. Its breakdown was delayed in sterilized soils, or when soil was at a lower moisture content, or when higher parthenin concentrations were applied to the soil. In addition, its breakdown could be promoted by higher temperatures and in soils with a higher potential cation exchange capacity and clay content. They concluded that certain physico-chemical and biological processes would govern the fate of parthenin in soil, and it is unlikely that parthenin would accumulate in soil over time. They speculated that if parthenin is to have a significant detrimental effect in the soil, it would require a high parthenium weed density to continuously produce parthenin, as well as a set of soil conditions that favour its persistence. In light of this, the long term ecological significance of parthenin release directly from parthenium weed plants is brought into question.

Table 3.3. Weeds that have been reported to be affected by parthenium weeds allelopathic capacity

Species	Scientific name	Produced by	How tested	Effect	Reference
Velvetleaf	*Abutilon multiflorum* Lam.	Shoot extract and shoot residue	Petri-dish Sand culture	Root growth reduced	Mersie and Singh, 1987
Barnyard grass	*Echinochloa crusgalli* L.	Leaf and flower extracts	Petri dish	Seedling growth reduced	Chen et al., 2011
Climbing buckwheat	*Fallopia convolvulus* L.	Leaf extract	Sand culture	Germination and seedling growth reduced	Adkins and Sowerby, 1996
Goose foot	*Chenopodium murale* L.	Parthenin extracted from leaves	Petri dish	Germination, seedlings growth and Chlorophyll content reduced	Batish et al., 2002
Amaranth	*Amaranthus viridis* L.				
Crofton Weed	*Ageratina adenophora* (Spreng) King	Leaf extract		Germination and seedling growth reduced	Maharjan et al., 2007
Mugwort	*Artemisia dubia* Wall ex. Besser				
Cinderella weed	*Synedrella nodiflora* L.	Living seedlings	Relay seeding	Shoot growth reduced and root stimulated	Belgeri et al., 2012
Dwarf amaranth	*Amaranthus marocarpus* Bench.				
Mikania vine	*Mikania micrantha* L.				
Plantain	*Plantago asiatica* L.	Leaf extract	Petri dish	Germination and seedling growth reduced	Hu et al., 2013
Hawkweed	*Youngia japonica* L.				
Canary grass	*Phalaris minor* Retz.	Leaf litter		Germination and seedling growth reduced	Aslam et al., 2014
Wild oat	*Avena fatua* L.	Leaf, root and whole plant extracts			

Evidence of the Role of Parthenin in Plant Litter

Belz et al. (2007) has studied the overall role of parthenin in the phytotoxicity of decomposing parthenium weed leaf litter. The natural release of parthenin during litter decomposition was stimulated by aqueous extraction and the responses of test species (*Ageratum conyzoides* L., *Echinochloa crus-galli* L., *Eragrostis curvula* (Schrad) Nees, *Eragrostis tef* (Zucc.) Trotter and *Lactuca sativa* L.) evaluated under laboratory conditions in dose-response bioassays. The test species differed considerably in their sensitivity to the litter extracts, with *A. conyzoides* being the most sensitive and *L. sativa* L. being the most resistant. Their results showed that the contribution of parthenin to the inhibition of germination by leaf litter was highly dependent upon its concentration and provided between 16 to 100% of overall phytotoxicity of the leaf litter extracts. They suggested that the release of parthenin during decomposition of leaf litter had the potential to play a significant role in the allelopathy of parthenium weed in field conditions.

Table 3.4. The major allelochemicals produced from different parts of parthenium weed

Plant part	Chemical family	Chemical constituents	Reference
Leaves	Sesquiterpenes lactones	Artecanin, balchanin, caffeic, coronopilin, costunolide, epoxyartemorin parthenin, p-coumeric acid, psedoguananolides, germacranolides and hymenin	Kohli et al., 2006
	Phenolic compounds	Anicic acid, ferulic acid, fumaric acid and vanicillic acid	Narwal et al., 2003
		p-counmaric acid, *p*-hydroxybenzoic acid gallic acid, 4-hydroxy-3-methoxy benzoic acid, m-counmaric acid, chlorogenic acid	Safdar et al., 2014
	Terpenoids	Myrcene, β-pinene, camphene, a-pinene, alcohol, and 1-phenylheptane	Chen et al., 2011
Stems	Phenolic compound	Caffeic acid, *p*-hydroxybenzoic acid, Gallic acid, m-counmaric acid and ferulic acid	Narwal et al., 2003
Roots	Phenolic compound	Caffeic acid, *p*-counmaric acid, ferulic acid, *p*-hydroxybenzoic acid, vanicillic acid gallic acid, m-counmaric acid, syringic acid and chlorogenic acid	Safdar et al., 2014
Flowers	Phenolic compound	Caffeic acid, *p*-counmaric acid and vanicillic acid	Narwal et al., 2003
	Terpenoids	Myrcene, ocimene, alcohol, camphene, 1-ocetene and 6,6- dimethyl-2-Methylenebicyclo [3.1.1]heptane	Chen et al., 2011
Pollen	Phenolic compound	Caffeic acid and *p*-counmaric acid	Kanchan and Jayachandra, 1980a
Aerial parts	Flavonoids	Aglycone flavanols	Shen et al., 1976
Whole plant	Phenolic compound	Caffiec acid, *p*-counmaric acid, anisic acid, *p*-hydroxybenzoic acid and vanillic acid	Das and Das, 1995
	Pseudoguaianolide	Parthenin, coronopilin and damsin	

Evidence of the Role of Phenolic Acids in Living Plants

Some of the phytochemicals that have been found in leaf and stem aqueous extracts of parthenium weed belong to the alkaloids, flavonoids, phenolic acids, glycosides, terpenoids, saponins, steroids and tannis (Narwal et al., 2003). Safder et al. (2014) claimed that the rhizospheric soil inhibition of maize germination and growth was caused by phenolic acids produced by parthenium weed roots. Flowers of parthenium weed have also been shown to contain more phenolic acids (vanillic acid, caffeic acid, chlorogenic acid and three isochlorogenic acids) than their leaves or roots (Ulrichs et al., 2010). In soil surrounding parthenium weed plants a higher concentration of water soluble phenolic acids has been found than was found in non-infested soils (Batish et al., 2005a).

3.4. Environmental Factor That Might Change the Allelopathic Activity

Temperature/Soil Moisture

Changes in the temperature, soil moisture and nutrient content, and microbial component of the soil can all have an influence upon allelopathic interactions between plants. Vaida et al. (2012) have reported extracts of European blueberry, common heather and feather moss to have a stronger inhibitory effect on the germination of Scots pine seeds at 20/29°C than at 18/27°C. Little has been undertaken on parthenium weed, but Safder et al. (2014) showed that plants growing in well-watered conditions were more allelopathic than those growing in dry areas. However, Shi (personal communication) has shown temperature in the range of 15 to 30°C did not to affect the allelopathic effect of parthenium weed litter on lettuce using the sandwich method.

Residue Burning

The allelopathic activity of unburnt as compared to burnt litter of parthenium weed have been studied on the growth of several crops (Singh et al., 2003). In general the aqueous extracts made from both unburnt and burnt were inhibitory to seedlings growth and dry weight production, however the burnt litter was more toxic than non-burnt litter, possibly due to its more alkaline nature. The activity of both kinds of extracts was attributed to the presence of phenolic acids (Singh et al., 2003). In another study Kumar and Kumar (2010) showed seeds of *Phaseolus mungo* not to germinate in aqueous extracts made from ash of burnt parthenium weed litter and Singh et al., (2004) has shown burnt litter of parthenium weed to be more inhibitory to the growth of radish and chickpea than unburnt residues.

Climate Change

It has been shown that plant height, leaf area, branch and tiller number, flowering and seed production of parthenium weed can be significantly enhanced under a changing climate of CO_2 enrichment (Khan et al., 2014; Shabbir et al., 2006 and Navie et al., 2005). As a C_3 plant parthenium weed was shown to grow better under an atmosphere of elevated CO_2. Wu et al. (1999) has shown that an extract from *Vulpia* litter can inhibit wheat seedlings growth more under 35°C than 25°C, which indicated that the allelopathic potential is greater in higher than lower temperature.

CONCLUSION

In conclusion, a large number of bioassays have been developed and used to try and establish what role if any allelopathy plays in the invasiveness of parthenium weed, a invasive plant that is now spreading at an alarming rate around the world and causing considerable economic and environmental damage. Most studies so far have concentrated on using a simple bioassay where an aqueous extract is applied to seed of a test species and then to assess the germination and seedling growth outcomes of that treatment. The main disadvantage of this approach is that, in most cases, much higher doses than would normally be produced in natural conditions are applied to the seeds, causing in many cases a false positive result. It is therefore essential that any aqueous extract bioassay undertaken should be followed by further approaches in the laboratory or glasshouse to confirm the presence of an allelopathic effect, but this is rarely done. Furthermore, it is very important to include in the bioassay at least one aqueous extract from a non allelopathic plant so it can be clearly demonstrated that the effect is due to an extracted allelopathic substance(s) and not to do with the myriad of other primary plant products that are also released during the extraction process, and again this is rarely done. A number of more biologically relevant bioassay approaches have been developed in recent years, and include the "relay seeding technique," " plant box" and "sandwich" methods, but they are also only rarely used in the study of parthenium weed. These assays have been successfully used to evaluate the natural release of allelopathic substances from plant roots and litter in a wide range of species. In several of these tests that have been undertaken on parthenium weed the allelopathic strength has been compared to other well-known allelopathic plants (studies by Akhtar et al., 2014; Channappagoudar et al., 2005; Shinwari et al., 2013a; b; Umer et al., 2010 etc.) and the results of such comparisons suggests that parthenium weed is only a moderately allelopathic plant. Although the full biochemistry of the allelopathic mechanism in parthenium weed is still incomplete, there is enough evidence to suggest that one main effect is produced by the release of phenolic acids from the plants roots, and for parthenin and phenolic acids to be released from its litter, mainly at the end of the growing season. It should be remembered that when growing conditions are ideal, very large numbers of parthenium weed seedlings will appear from its uncommonly large seed bank and immediately start to release allelochemicals. Even if the allelopathic nature of each plant is only moderate, very soon the landscape will become devoid of any other plant type as is commonly seen in parthenium weed infested regions and a complete monoculture of the weed becomes established. So, under these conditions it can be speculated that the moderate level of allelopathy of each plant is creating an extremely strong allelopathic population which presumably will be playing a major role in the future invasiveness of parthenium weed, but at the population level.

REFERENCES

A National Program of the States, Territories and the Australian Government (2009). Fact Sheet Declared Class 2 Pest Plant Parthenium Weed Parthenium hysterophorus. The State of Queensland, Department of Employment, Economic Development and Innovation.

Adamson, DC (1996). Introducing dynamic considerations when economically evaluating weeds. The University of Queensland, Brisbane. MSc Thesis.

Adkins, SW and Navie, SC (2006). Parthenium weed: a potential weed for agro-ecosystems in Pakistan. *Pakistan Journal of Weed Science Research* 12 (1/2): 19-36.

Adkins, SW, Navie, SC and Dhileepan, K (2005). Parthenium weed in Australia: Research progress and prospects. In *Proceedings of the 2nd International Conference on Parthenium Management*, Ed. T.V.R. Prasad, pp.11-27. University of Agricultural Sciences, Bangalore, India.

Adkins, SW and Sowerby, MS (1996). Allelopathic potential of the weed, *Parthenium hysterophorous* L., in Australia. *Plant Protection Quarterly* 11(1): 20-23.

Adkins, SW, O'Donnell, C, Khan, N, Nguyen, TL, Shabbir, A and Dhileepan, K (2010). Parthenium weed (*Parthenium hysterophorus* L.) research in Australia: New management possibilities. In *Proceedings of the 17th Australasian Weeds Conference*, ed. S.M. Zydenbos, p.120-123. Christchurch, New Zealand. New Zealand Plant Protection, 26-30 September, 2010.

Adkins, SW, O'Donnell, C, Khan, N, Nguyen, TL, Shabbir, A and Dhileepan, K (2010). Parthenium weed (*Parthenium hysterophorus* L.) research in Australia: New management possibilities. In *Proceedings of the 17th Australasian Weeds Conference*, ed. S.M. Zydenbos, p.120-123. Christchurch, New Zealand. New Zealand Plant Protection, 26-30 September, 2010.

Adkins, SW, Peters, NCB, Paterson, MF and Navie, SC (2003). Germination stimulation of weed species by smoke. In *The Biology of Seeds: Recent Research Advances. Proceedings of the Seventh International Workshop on Seeds, Salamanca, Spain, 2002.*, pp. 413-420.

Adkins, SW and Shabbir, A (2014). Biology, ecology and management of the invasive parthenium weed (*Parthenium hysterophorus* L.). Pest Management Science. doi: 10.1002/ps.3708.

Agarwal, C. and Anand, A. (1992). Ecological studies on allelopathic potential of *Parthenium hysterophorus* L. in relation to *Phaseolus aureus* L. and *Triticum aestivum* L. In Proceedings of the First National Symposium on Allelopathy in Agroecosystems, eds. P. Tauro and S.S. Narwal, p.64-65. Haryana Agricultural University, Hisar, 12-14 February, 1992.

Ahmed, MN, Rao, PR and Mahendar, M (1988). Haematological observations in experimental partheniosis in buffalo calves. *Indian Veterinary Journal* 65(11): 972-974.

Akhtar, S, Bangash, N, Asghar, R, Munir, M and Khalid, N (2014). Allelopathic assessment of selected invasive species of Pakistan. *Pakistan Journal of Botany* 46(5): 1709-1713.

Aneja, KR, Dhawan, SR and Sharma, AB (1991). Deadly weed, *parthenium hysterophorus* L. - and its distribution. *Indian Journal of Weed Science* 23:14-18.

Anjum, A, Hussain, U, Yousaf, Z, Khan, F and Umer, A (2010). Evaluation of allelopathic action of some selected medicinal plant on lettuce seeds by using sandwich method. *Journal of Medicinal Plants Research* 4(7): 536-543.

Anonymous (2003). Incursion of parthenium weed (*Parthenium hysterophorus*) in Pupua New Guinea. Plant Protection Service, Secretariat of the Pacific Community PNG.

Aslam, F, Khaliq, A, Matloob, A, Abbas, RN, Hussain, S and Rasul, F (2014). Differential allelopathic activity of *Parthenium hysterophorus* L. against canary grass and wild oat. *The Journal of Animal and Plant Sciences* 24(1): 234-244.

Babu, GP, Hooda, V, Audiseshamma, K and Paramageetham C (2014). Allelopathic effects of some weeds on germination and growth of *Vigna mungo* (L.) Hepper. *International Journal of Current Microbiology and Applied Sciences* 3(4): 122-128.

Batish, DR, Singh, HP, Kohli, RK, Saxena, DB and Kaur, S (2002a). Allelopathic effects of parthenin against two weedy species, *Avena fatua* and *Bidens pilosa*. *Environmental and Experimental Botany* 47: 149-155.

Batish, DR, Singh, HP, Pandher, JK, Arora, V and Kohli, RK (2002b). Phytotoxic effect of parthenium residues on the selected soil properties and growth of chickpea and radish. *Weed Biology and Management* 2: 73-78.

Batish, DR, Singh, HP, Pandher, JK, Kohli, RK (2005a). Invasive potential of the weed Parthenium hysterophorus - the role of allelopathy. In: Plant Protection and Plant Health in Europe: Introduction and Spread of Invasive Species, Held at Humboldt University, Berlin, Germany, pp. 9-11.

Batish, DR, Singh, HP, Pandher, JK and Kohli, RK (2005b). Allelopathic interference of *Parthenium hysterophorus* residues in soil. *Allelopathy Journal* 15(2): 267-274.

Belgeri, A, Navie, S and Adkins, SW (2012). Screening parthenium weed (*Parthenium hysterophorus* L.) seedlings for their allelopathic potential. *Pakistani Journal of Weed Science Research* 18: 727-731.

Belgeri, A. (2013). Assessing the impact of parthenium weed (*Parthenium hysterophorus* L.) invasion and its management upon grassland community composition. The University of Queensland. PhD thesis.

Belz, RG, Vander, LM, Reinhardt, CF and Hurle, K (2009). Soil degradation of parthenin does it contradict the role of allelopathy in the invasive weed *Parthenium hysterophorus* L. *Journal of Chemical Ecology* 35(9): 1137-1150.

Belz, RG, Reinhardt, CF and Foxcroft, LC (2007). Residue allelopathy in *Parthenium hysterophorus* L – Does parthenin play a leading role? *Crop Protection* 26: 237-245.

Bahtt, BP, Chauhan, DS and Todaria, NP (1994). Effect of weed leachates on germination and radicle extension of some food crop. *Indian Journal of Plant Physiology* 37(3): 177-179.

Bhawana, J, Sarika, Pandey, N and Rao, PB (2009). Allelopathic effect of weed species extraction germination, growth and biochemical aspects in different varieties of wheat (*Triticum aestivum* L.). *Indian Journal Agricultural Resources* 49: 79-87.

Bhowmik, PC and Doll, JD (1982). Corn and soybean response to allelopathic effects of weed and crop residues. *Agronomy Journal* 74: 601-606.

Bisikwa, J (2010). Parthenium weed distribution and management in Uganda. In *Programme and Abstracts, First IOBC International Workshop on Biological Control and Management of Parthenium Weed*, Nairobi, p. 26.

Bousquet, J, Reid, J and Van, WC (1997). Allergic rhinitis management pocket. *Allergy* 63 (8):990-996.

Bowen (2001). A study of the extent of alien species invasion into the Lubombo Conservancy, Swaziland. http://www.questoverseas.com/active africa /documents /AstudyOfTheExtentSwaziland.doc.

Channappagoudar, BB, Jalageriand, BR and Biradar NR (2005). Allelopathic effect of aqueous extracts of weed species on germination and seedling growth of some crops. *Karnataka Journal Agricultural Science* 18(4): 916-920.

Chen, YB, Wang JX, Wu XH, Sun JA and Yang N (2011). Allelopathic effects of *Parthenium hysterophorus* L. volatiles and its chemical components. *Allelopathy Journal* 27: 217-223.

Chippendale, JF and Panetta, FD (1994). The cost of parthenium weed to the Queensland cattle industry. *Plant Protection Quarterly* 9: 73-76.

Chou, CH and Muller, CH (1972). Allelopathic mechanisms of *Arctostaphylos glandulosa vat. zacaensis*. *American Midland Naturalist* 88: 324-347.

Clark, K, Lotter, W and Brill, D (2010). Status of *Parthenium hysterophorus* L. in Tanzania. In *Program and Abstracts, First IOBC International Workshop on Biological Control and Management of Parthenium Weed*, Nairobi 1-5 November, 2010, p. 27.

Dale, IJ (1981). Parthenium weed in the Americas. *Australian Weeds* 1 (1): 8-14.

Das, B and Das R (1995). Chemical investigation in *Parthenium hysterophorus* – An allelopathic plant. *Allelopathiy Journal* 2(1), 99-104.

Dayama, DP (1986). Allelopathic effect of *Parthenium hysterophorus* L. on growth, nodulation and nitrogen content of Leucaena leucocephala. *Leucaena Research Reports* 7: 36-37.

Demissie, AG, Ashenafi A, Arega, A, Etenash U, Kebede, A and Tigist A (2013). Effect of parthenium hysterophorus L. on germination and elongation of onion (*Allium cepa*) and Bean (*Phaseolus vulgaris*). *Research Journal of Chemical and Environmental Sciences*, 1(2): 17-21.

Devi, MT and Raghavendra, AS (1994). Formation of photorespiratory glycollate or glyoxylate in leaf discs of C3-C4 intermediates: reduced sensitivity to external bicarbonate. *Photosynthetica* 30: 455-463.

Dhawan, P and Dhawan, SR (1995). Allelochemic effect of *Parthenium hysterophorus* L. on the germination behaviour of some agricultural crops. *Flora and Fauna* 1: 59-60.

Dhileepan, K (2007). Biological control of parthenium (*Parthenium hysterophorus*) in Australian rangeland translates to improved grass production. *Weed Science* 55: 497-501.

Dongre, PN, Chaubey, KS and Singh, AK (2010). Effects of leaf extracts of weeds on growth and yield of green gram. *Allelopathy Journal* 25 (1): 213-220.

Dongre, PN and Singh, AK (2007). Inhibitory effects of weeds on growth of wheat seedlings. *Allelopathy Journal* 20(2): 387-394.

EPPO, 2014. PQR database. Paris, France: European and Mediterranean Plant Protection Organization. http://www.eppo.int/DATABASES/pqr/pqr.htm.

Etana, A, Kelbessa, E and Soromessa, T (2015). Impact of *Parthenium hysterophorus* L. (Asteraceae) on soil chemical properties and its distribution in a reserve area: A case study in Awash National Park (ANP), Ethiopia. *Journal of Soil Science and Environmental Management* 6(5): 116-124.

Evans, HC (1997). *Parthenium hysterophorous*: a review of its weed status and the possibilities for biological control. *Biocontrol News and Information* 18: 389-398.

Everist, SL (1976). Parthenium weed. *Queensland Agricultural Journal* 102 (2): (cover page).

Friend, E (1983). Compositae (daisy family); tribe Heliantheae; subtribe Melampodiinae. Queensland weed seeds. Queensland Department of Primary Industries, Brisbane, p. 33.

Fuente, JR, Uriburu, ML, Burton, G and Sosa, VE (2000). Sesquiterpene lactone variability in *Parthenium hysterophorus* L. *Phytochemistry* 55 (7): 769-772.

Fujii, Y (1992). The potential biological control of paddy weeds with Allelopathy (allelopathic effects of some rice varieties). In: Proceedings of International Symposium. Biological Control and Integrated Management of Paddy and Aquatic Weeds in Asia, National Agriculture Resource Centre, Tsukuba, Japan, Food and Fertilisation Technical Centre. for the Asian and Pacific Region, China, pp. 305–320.

Fujii, Y, Matsuyama, M, Hiradate, S, and Shimozawa, H (2005). Dish pack method: A new bioassay for volatile allelopathy. In: (Eds.): JDI Harper, M An, H Wu and JH Kent, *Proceedings of the 4th World Congress on Allelopathy, "Establishing the Scientific Base,"* Fourth World Congress on Allelopathy, Wagga Wagga, New South Wales.

Fujii, Y, Syeda, SP, Mohammad, MP, Yoshio, O and Osamu, I (2003). Screening of 239 medicinal plant species for allelopathic activity using the sandwich method. *Weed Biology and Management* 3: 233-241.

Gil Rhee, D (2004). Alien plants in Korea, *National Institute of Environmental Research (NIER)*, viewed 12/6/2010 accessed <http://www.nier.go.kr >.

Gunaseelan, VN (1998). Impact of anaerobic digestion of inhibition potential of Parthenium soids. *Biomass Bioenergy* 14:179–184.

Gupta, A (1996). Allelopathic influence of *Parthenium hysterophorus* L. over Amaranthus gangeticus and overcoming toxicity. *Flora and Fauna* 2 (1): 57-58.

Gupta, A (2000). Allelopathic potential of root extracts of *Parthenium hysterophorus* L. *Allelopathic Journal* 7(1): 105-108.

Gupta, S and Narayan, R (2010). Effects of applied leaf biomass of *Parthenium hysterophorus*, *Cassia obtusifolia* and *Achyranthes aspera* on seed germination and seedling growth of wheat and pea. *Allelopathy Journal* 26(1): 59-70.

Guzman, CD (1988). Allelopathic effects of seven weed species in pumpkin (*Cucurbita moschata*) under greenhouse conditions. *Journal of Agriculture, University of Puerto Rico* 72(3): 491-493.

Haig, T (2008). Allelochemicals in Plants. Allelopathy in Sustainable Agriculture and Forestry, Springer New York.

Hammerton, JL (1981). Weed problems and weed control in the Commonwealth Caribbean. *Tropical Pest Management* 27(3): 379-387.

Haseler, WH (1976). *Parthenium hysterophorous* L. in Australia. *Pesticides Articles and News Summaries* 24: 325-332.

Hassan, G and Khan, H (2010). Distribution, socio-economic and environmental impact of Parthenium weed in Peshawar valley, northwest Pakistan. In Scientific Presentations on Parthenium, 3rd International Conference on Parthenium. Eds. R.D. Gautam, G.K. Mahapatro, Shashi Bhalla, K. Shankarganesh, Sudhida Gautam, Kaushal Verma and H.S. Gaur, pp. 26-27. December 8-10, Indian Agricultural Research Institute, New Delhi 110012, India.

Hossain, I, Karim, R, Nag, BL, Adkins, SW and Mondal MRI (2015). Nationwide survey of parthenium weed infestation: A major threat to food security in Bangladesh. *In proceeding of the 25th Asian-Pacific Weed Science Society Conference on Weed Science for Sustainable Agiculture, Environment and Biodiversity, Hyderabad, India.*

Houghton, JT, Ding, Y, Griggs, DJ, Noguer, M, van der Linden, PJ and Xiaosu, D (2001). Climate change 2001: The Scientific Basis. Contribution of Working Group I to the Third Assessment Report of the Intergovernmental Panel on Climate Change. Cambridge University Press, Cambridge, United Kingdom and New York.

Hu, G, Zhang, Z and Hu, B (2013). Effect of aqueous leaf extract of *Parthenium hysterophorus* L. on the germination and shoot growth of two native species. *Advanced Materials Research* 726-731: 4348-4351.

Inderjit and Dakshini, KMM (1995). On laboratory bioassays in allelopathy. *Botanical Review* (Lancaster) 61: 4-9.

Jayasurya, AHM (2005). Parthenium weed – status and management in Sri Lanka. In T Prasad, N HV andDRea (eds) (eds), *Second international conference on parthenium management*, University of Agricultural Sciences, Bangalore, India, pp. 36–43.

Joel, DM and Liston, A (1986). New adventives weeds in Israel. *Israel Journal of Botany* 35: 215-223.

Kanchan, SD and Jayachandra (1979). Allelopathic effects *of Parthenium hysterophorus* L.1. Exudation of inhibitors through roots. *Plant Soil* 34, 27-35.

Kanchan SD and Jayachandra (1980a). Allelopathic effects of *Parthenium hysterophorus* L. 2. Leaching of inhibitors from aerial vegetative parts. *Plant and Soil* 55:61–66.

Kanchan, SD and Jayachandra (1980b). Pollen allelopathy – a new phenomenon. *New Phytologist* 84: 739-746.

Karl, TR and Trenberth, KE (2003). Morden global climate change. *Science* 302: 1719-1723.

Karim, SM and Rezaul (2013). Parthenium invasion in Malaysia: a possibility of ecological disaster.Http://fiat.umk.edu.my/bm/images/gallery/article_2013_parthenium_invasion_in_malaysia.pdf.

Kato-Noguchi H and Peters RJ (2013). The role of momilactones in rice allelopathy. *Journal of Chemical Ecology* 39(2): 175-185.

Kbosla, SN and Sobti, SN (1981). Effective control of *parthenium hysterophorus* Linn. *Pesticides (India)* 15 (4): 18-19.

Khan, N (2011). Long term, sustainable management of parthenium weed (*Parthenium hysterophorus* L.) using suppressive plants. The University of Queensland. PhD thesis.

Khan, N, Shabbir, A, George, D, Hassan, G and Adkins, SW (2014). Suppressive fodder plants as part of an integrated management program for *parthenium hysterophorus* L. *Field Crops Research* 156: 172-179.

Khosla, S and Sobti, S (1979). Parthenium – a national health hazard, its control and utility – a review. *Pesticides* 13: 121–127.

Kilian, N, Peter, H and Ali, HM (2002). New and noteworthy records for the flora of Yemen, chiefly of Hadhramout and Al-Mahra. *Willdenowia* 32: 239-269.

Kohli, RK, Rani, D (1992). Identification and bio efficacy of soil chemicals of Parthenium. In: Tauro, P., Narwal, S.S. (Eds.), Proceedings of the 1st National Symposium on Allelopathy in Agroecosystems, Hisar, India, February 1992. Hisar; Haryana Agricultural University, pp. 196-198.

Kohli, RK, Batish, DR, Singh, HP and Sogra, S (2006). Status, invasiveness and environmental threats of three tropical American invasive weeds (*Parthenium hysterophorus* L., *Ageratum conyzoides* L., *Lantana camara* L.) in India. *Biological Invasions* 8: 1501-1510.

Lakshmi, C and Srinivas, C (2007). Parthenium: A wide angle view. *Indian Journal Dermatology Venereology and Leprology* 73(5): 296-306.

Lewis, WH, Dixit, AB and Wedner, HJ (1988). Reproductive biology of *Parthenium hysterophorus* (Asteraceae). *Journal of Palynology* 24: 73-82.

MacKee, HS (1994). Catalogue des plantes introduites et cultivées en Nouvelle-Calédonie. Muséum National d'Histoire Naturelle, Paris.

Madhu, M, Nanjappa, HV and Ramachandrappa, BK (1995). Allelopathic effect of weeds on crops. *Mysore Journal of Agricultural Sciences* 29 (2): 106-112.

Mahadevappa, M (2009). Parthenium: Insight into its Menace and Management. Stadium Press (India), New Delhi.

Maharjan, S, Shrestha, BB, Jha, PK (2007). Allelopathic effects of aqueous extract of leaves of *Parthenium hysterophorus* L. on seed germination and seedling growth of some cultivated and wild herbaceous species. *Scientific World Journal* 5: 33-39.

Martinez, FD (2007). Genes, environments, development and asthma: a reappraisal. *European Respiration Journal* 29 (1): 179–184.

McConnachie, AJ, Strathie, LW, Mersie, W, Gebrehiwot, L, Zewdie, K, Abdurehim, A, Abrha, B, Araya, T, Asaregew, F, Assefa, F, Gebre-Tsadik, R, Nigatu, L, Tadesse, B and Tana, T (2011). Current and potential geographical distribution of the invasive plant *Parthenium hysterophorus* (Asteraceae) in eastern and southern Africa. *Weed Research* 51(1): 71-84.

McFadyen, RE (1992) Biological control against parthenium weed in Australia. *Crop Protection* 24: 400–407.

McFadyen, RE (1995) Parthenium weed and human health in Queensland. *Australia Farm Physician* 24:1455-1459.

Medhin, BG (1992). *Parthenium hysterophorus*, new weed problem in Ethiopia. *FAO Plant Protection Bulletin* 40: 49.

Mersie, W and Singh, M (1987). Allelopathic effect of parthenium (*Parthenium hysterophorus* L.) extract and residueon some agronomic crops and weeds. *Journal of Chemical Ecology* 13: 1739-1747.

Mersie, W and Singh M (1988). Effects of phenolic acids and ragweed parthenium (*Parthenium hysterophorus*) Extracts on Tomato (*Lycopersicon esculentum*) Growth and Nutrient and Chlorophyll content. *Weed Science* 36: 278-281.

Miranda, CASF, Cardoso MG, Carvalho, MLM, Figueiredo, ACS, Nelson, DL, Oliveira, CM, Gomes, MS, Andrade, J, Souza, JA and Albuquerque, LRM (2014). Chemical composition and allelopathic activity of *Parthenium hysterophorus* and *Ambrosia polystachya* weeds essential oils. *American Journal of Plant Sciences* 5(9): 1248-1257.

Mishra, KK (1991). Parthenium hysterophorus LINN. – a new record for Nepal. *Journal of the Bombay Natural History Society* 88(3): 466-467.

Mito, T and Uesugi T (2004). Invasive alien species in Japan: The status quo and the new regulation for prevention of their adverse effects. *Global Environmental Research* 2: 171-179.

Motooka, P, Castro, L, Nelson, D, Nagai, N and Ching, L (2003). Weeds of Hawaii's pastures and natural areas: An identification and management guide. College of Tropical Agriculture and Human Resources, University of Hawaii Manoa, Honolulu.

Msafiri, CJ, Tarimo, MT and Ndakidemi, PA (2013). Allelopathic effects of *Parthenium hysterophorus* on seed germination, seedling growth, fresh and dry mass production of *Alysicurpus glumaceae and Chloris gayana. American Journal of Research Communication* 1(11): 190-205.

Nandhakishore, T and Pasricha, JS (1994). Pattern of cross-sensitivity between 4 Compositae plants, *Parthenium hysterophorus*, *Xanthium strumarium*, *Helianthus annuus* and *Chrysanthemum coronarium*, in Indian patients. *Contact Dermatitis* 30:162-167.

Narasimhan, TR, Ananth, M, Swamy, MN and Rao, PVS (1977). Toxicity of *Parthenium hysterophorus* L.. *Current Science* 46: 15-16.

Narwal, SS, Palaniraj, R, Sati, SC, Kadian, HS and Dahiya, DS (2003). Allelopathic plants: 8. *Parthenium hysterophorus* L. *Allelopathy Journal* 11(2): 151-170.

Nath, R (1981). Note on the effect of *Parthenium* extract on seed germination and seedling growth in crops. *Indian Journal of Agricultural Science* 51: 601-603.

Nath, R (1988). *Parthenium hysterophorus* L. – a review. *Agriculture Reviews* 9(4): 171-179.

Navarez, DC and Olofsdotter, M (1996). Relay seeding technique for screening allelopathic rice (*Oryza sativa* L.). pp. 1285-1290 in Proc. 2nd International Weed Control Congress, Copenhagen, Denmark.

Navie, SC (2002). *The Biology of* Parthenium hysterophorous *L. in Australia.* PhD Dissertation. The School of Land Crop and Food Sciences, The University of Queensland, Brisbane, Australia.

Navie, SC, McFadyen, RE, Panetta, FD and Adkins, SW (1996). A comparison of the growth and phenology of two introduced biotypes of *Parthenium hysterophorus*. In *Proceedings of the Eleventh Australian Weeds Conference*, Ed. Shepherd, RHC, pp. 313-316. Weed Science Society of Victoria, Melbourne.

Navie, SC, Mc Fadyen, RE, Panetta, FD and Adkins, SW (2005). The effect of CO_2 enrichment on the growth of a C3 weed (*Parthenium hysterophorus* L.) and its competitive interection with a C4 grass (*Cenchrus ciliaris* L.). *Plant Protection Quartely* 20: 61-66.

Netsere, A and Mendesil, E (2011). Allelopathic effects of *Parthenium hysterophorus* L. aqueous extracts on soybean (*Glycine max* L.) and haricot bean (*Phaseolus vulgaris* L.) seed germination, shoot and root growth and dry matter production. *Journal of Applied Botany and Food Quality-Angewandte Botanik* 84: 219-222.

Nguyen, TLT (2011). The invasive potential of parthenium weed (*Parthenium hysterophorus* L.) in Australia. The University of Queensland. PhD thesis.

Ni, GY, Zhao, P, Huang, QQ, Hou, YP, Zhou, CM, Cao, QP and Peng, SL (2012). Exploring the Novel Weapons Hypothesis with invasive plant species in China. *Allelopathy Journal* 29(2): 199-214.

Nigatu, L, Hassen, A., Sharma, J.J and Adkins, SW (2010). Impact of *Parthenium hysterophorus* L. on grazing land communities in North-Eastern Ethiopia. *Weed Biology and Management* 10: 143-152.

Njoroge, JM (1986). New weeds in Kenya coffee. A short communication. *Kenya Coffee* 51: 333–335.

Njoroge, JM (1991). Tolerance of *Bidens pilosa* L. and *Parthenium hysterophorus* L. to paraquat (Gramoxone) in Kenya coffee. *Kenya Coffee* 56: 999-1001.

Noor, K, Khan, EA, Baloch, MS, Khan, MA, Awan, IU, Sadiq, M and Aslam M (2012). Allelopathic effect of congress grass on weeds and yield of wheat. *Pakistan Journal of Weed Science Research* 18(3): 307-318.

Pandey, DK (1994). Inhibition of salvinia (Salvinia molesta Mitchell) by Parthenium (*Parthenium hysterophorus* L.) II. Relative effect of flower, leaf, stems, and root residue on salvinia and paddy. *Journal Chemical Ecology* 20: 3123-3131.

Pandey, HP, Raza, AK and Chauhan, SK (2011). Assessment of allelopathic aggression of *Parthenium hysterophorus* L. on seed germination and seedling growth of some important cereals. *Trends in Biosciences* 4(1): 68-70.

Parker, C (1992). Weeds of Bhutan. Royal Government of Bhutan: Bhutan.

Parsons, WT and Cuthbertson, EG (2001). *Noxious Weeds of Australia.* Second edition CSIRO publishing. Collingwood, Vic. Australia.

Parthasarathi, T, Suganya V and Sivakumar R (2012). Allelopathic effect of aqueous leaf extract of parthenium hysterophorus L. on seed germination and seedling growth in greengram, blackgram and groundnut. *Madras Agricultureal Journal* 99(7-9):514-517.

Peng, CI, Hu, LA and Kao, MT (1988). Unwelcome naturalisation of *Parthenium hysterophorus* (Asteraceae) in Taiwan. *Journal of Taiwan Museum* 41 (2): 95-101.

Picman, J and Picman, AK (1984). Autotoxicity in *Parthenium hysterophorus* and its possible role in control of germination. *Biochemical Systematics and Ecology* 12: 287-292.

Putnam, AR and Duke WB (1974). Biological suppression of weeds: Evidence for allelopathy in accessions of cucumber. *Science* 185: 370-372.

Putnam, AR and Tang, CS (1986). Alleopathy: state of the science, in AR Putnam and CS Tang (eds), *The Science of Allelopathy*, Wiley and Sons, New York, pp 1-19.

Queensland Government (2011). *Fact Sheet – Weeds of Australia Biosecurity Queensland Edition, Parthenium weed Parthenium hysterophorus* L. http://keyserver.lucidcentral.org /weeds/data/03030800-0b07-490a-8d04-0605030c0f01/media/Html/Parthenium_hystero phorus.htm.

Rajendrudu, G and Rama VSD(1990). C3-like carbon isotope discrimination in C3–C4 intermediate *Alternenthera* and *Parthenium* species. *Current Science* 59: 377–379.

Ramaswarmi, PP (1997). Potential uses of parthenium weed. In Mahadevappa, M and Patil, VC Eds. In Proceedings of the 1st International Conference on Parthenium Management. University of Agriculture Sciences: Dharwad, India, pp. 13-17.

Rao, PA and Rajendrudu, D (1989). Net photosynthetic rate in relation to leaf anatomical characteristics of C3, C3-C4 and C4 dicotyledons. *Plant Science* 99: 529-537.

Rao, RS (1956). Parthenium: a new record for India. *Journal of Bombay Natural History* 54: 218-220.

Razaq, ZA, Vahidy, AA and Ali, SI (1994). Chromosome numbers in Compositae from Pakistan. *Annals of the Missouri Botanical Garden* 81(4): 800-808.

Reinhardt, C, Van der Laan, M, Belz, RG, Kurle, K and Foxcroft, L (2006). Production dynamics of the allelochemical parthenin in leaves of *Parthenium hysterophorus* L. *Journal of Plant Diseases and Protection* Special Issue XX: 427-433.

Rice, EL (1984). Allelopathy. Academic Press, 2nd Edition, New York.

Rollins, RC (1950). The guayule rubber plant and its relations. *Contribution from the Gray Herbarium* 172: 1-72.

Roy, DC and Shaik, MM (2013). Toxicology, Phytochemistry, Bioactive compounds and Pharmacology of Parthenium hysterophorus. *Journal of Medicinal Plants Studies* 1(3): 126-141.

Safdar, ME, Tanveer, A, Khaliq, A and Naeem, MS (2014). Allelopathic action of parthenium and its rhizospheric soil on maize as influenced by growing conditions. *Planta Daninha, Vicosa-MG* 32(2): 243-253.

Sajjan, AS and Pawar, KN (2005). Studies on allelopathic effects of *Parthenium hysterophorus* on some crop species. *Agricultural Science Digest* 25(3): 166-169.

Satsangi, GP, Shrivastava, JN, Kumari, S and Gangwar, N (2002). Allelopathic responses of *Parthenium hysterophorus* L. on germination and seedling vigour in *Sorghum vulgare* L. and *Cajanus cajan* L. and control measures for the weed. Thirteenth Australian Weeds Conference.

Shabbir, A and Bajwa, R (2006). Distribution of parthenium weed (*Parthenium hysterophorus* L.), an alien inrasive weed species threatening the biodiversity of Islamabad. *Weed Biology Management* 6: 89-95.

Shabbir, A (2012). Towards the improved management of parthenium weed: complementing biological control with plant suppression. The University of Queensland. PhD thesis.

Shen, MC, Rodriguez, E, Kerr, KTJ and Marby, K (1976). Flavonoids of four species of Parthenium (compositae). *Phytochemistry* 15: 1045-1047.

Shi, B (2012). Screening parthenium weed (*Parthenium hysterophorus* L.) for allelopathic potential against fodder grasses using the "plant-box" and "sandwich-box" methods. A report submitted as a requirement for the degree of Masters in Plant Protection Studies at the University of Queensland.

Shinwari, MI, Shinwari, MI and Fujii, Y (2013a). Allelopathic activity of medicinal plants and weeds from Pakistan. *Allelopathy Journal* 32(2): 223-232.

Shinwari, MI, Shinwari, MI and Fujii, Y (2013b). Allelopathic evaluation of shared invasive plants and weeds of Pakistan and Japan for environmental risk assessment. *Parkistan Journal Botany* 45(S1): 467-474.

Shrestha, BB, Shabbir, A and Adkins, SW (2015). *Parthenium hysterophorus* in Nepal: a review of its weed status and possibilities for management. *Weed Research* 55, 132-144.

Singh, HP, Batish, DR, Pandher, JK and Kohli RK (2003). Assessment of allelopathic properties of *Parthenium hysterophorus* residues. *Agricultural Ecosystem Environment* 95(2-3): 537-541.

Singh, HP, Batish, DR, Pandher, JK and Kohli, RK (2005). Phytotoxic effects of *Parthenium hysterophorus* residues on three *Brassica* species. *Weed Biology and Management* 5 (3): 105-109.

Singh, SP and Sangeeta (1991). Allelopathic potential of *Parthenium hysterophorus* L. *Journal Agronomy and Crop Science* 167: 201-206.

Srivastava, JN, Shukla, JP, Srivastava, RC (1985). Effect of *Parthenium hysterophorus* Linn. extract on the seed germination and seedling growth of barley, pea and wheat. *Actatic Botanica Indician* 13: 194-197.

Swain, D, Paroha, S, Singh, M and Subudhi, HN (2012). Evaluation of allelopathic effect of *Echinochloa colona* weed on rice (*Oryza sativa* L. 'Vandana'). *Journal of Environmental Biology* 33(5): 881-889.

Swaminathan, C, Rai, RSV and Suresh, KK (1990). Allelopathic effects of *Parthenium hysterophorus* on germination and growth of a few multi-purpose trees and arable crops. *International Tree Crops Journal* 6 (2-3): 143-150.

Tadelle, T (2002). Allelopathic Effects of *Parthenium hysterophorus* Extracts on Seed Germination and Seedling Growth of *Eragrostis tef*. *Journal of Agronomy and Crop Science* 188: 306-310.

Tamado, T, Ohlander, L and Milberg, P (2002a). Interference by the weed *Parthenium hysterophorus* L. with grain sorghum: influence of weed density and duration of competition. *International Journal of Pest Management* 48(3): 183-188.

Tamado, T, Schutz, W and Milberg, P (2002b). Germination ecology of the weed *Parthenium hysterophorus* in eastern Ethiopia. *Annals of Applied Biology* 140: 263-270.

Tamer, M, Sanjay, G and El-Keblawy, A (2015). *Parthenium hysterophorus* and *Bidens pilosa*, two new records to the invasive weed flora of the United Arab Emirates. *Journal on New Biological Reports* 4(1): 26-32.

Tanveer, A, Khaliq, A, Ali HH, Mahajan, G and Chauhan, BS (2015). Interference and management of parthenium: The world's most important invasive weed. *Crop Protection* 68: 49-59.

Tefera, T (2002). Allelopathic effects of *Parthenium hysterophorus* extracts on seed germination and seedling growth of Eragrostis tef. *Journal of Agronomy and Crop Science* 188 (5): 306-310.

Thorp, JR and Lynch, R (2000). The determination of Weeds of National Significance. National Weeds Strategy Executive Committee, Launceston.

Timsina B, Shrestha BB, Rokaya MB, Munzbergova Z (2011). Impact of *Parthenium hysterophorus* L. invasion on plant species composition and soil properties of grassland communities in Nepal. *Flora* 206 (3): 233-240.

Towers, GH, Mitcheu, N, Rodriguez, JC Bennett, E and Rao, SPVA (1977). Biology and chemistry of *Parthenium hysterophorus* L. a problematic weed in India. *Journal of Scientific and Industrial Research* 36: 672-684.

Tudor, GD, Ford, AL, Armstrong, TR and Bromage, EK (1982). Taints in meat from sheep grazing *Parthenium hysterophorus. Australian Journal of Experimental Agriculture* 22: 43-46.

Ulrichs, C, Buettner, C, Wiesner, M, Roth, I and Blievernicht, A (2010). Eco-Physiological Aspects of Parthenium weed (*Parthenium hysterophorus* L.) Management in Ethiopia. XXVIII International Horticultural Congress on Science and Horticulture for People (IHC2010): International Symposium on Advances in Ornamentals, Landscape and Urban.

Umer, A, Yousaf, Z, Khan, F, Hussain, U, Anjum, A, Nayyab, Q and Younas A (2010). Evaluation of allelopathic potential of some selected medicinal species. *African Journal of Biotechnology* 9(37): 6194-6206.

USDA-ARS (2012). Germplasm Resources Information Network (GRIN). Beltsville, Maryland, USA: National Germplasm Resources Laboratory. http://www.ars-grin.gov.

Vaida, S, Ligita, B and Remigijus, O (2012). Allelopathic impact of some dominants in clean cuttings of Scots pine forest under climate change conditions. *EKOLOGIJA* 58(2): 59-64.

Valliappan, K and Towers, GHN (1988). Allelopathic effect of root exudates from the obnoxious weed, *Parthenium hysterophorus. Indian Journal of Weed Science* 20 (1): 18-22.

Venkataiah, B, Ramesh, C and Ravindranath, N (2003). Charminarone, a seco-pseudoguaianolide from *Parthenium hysterophorus. Phytochemistry* 63: 383–386.

Wakjira, M. (2009). Allelopathic effects of Parthenium hysterophorus L. on germination and growth of onion. *Allelopathy Journal* 24(2): 351-362.

Wakjira, M, Berecha, G and Tulu, S (2009). Allelopathic effects of an invasive alien weed Parthenium hysterophorus L. compost on lettuce germination and growth. *African Journal of Agricultural Research* 4 (11): 1325-1330.

Wang, Q, Ruan, X, Li, ZH and Pan, CD (2006). Autotoxicity of plants and research of coniferous forest autotoxicity. *Scientia. Silvae. Sinicae* 43: 134-142.

Warren, R and Heymann, MD (2006). Hand dermatitis, dialogues in dermatology. *Journal of American Academy of Dermatology* 54(6): 1078-1080.

Weston, LA, Harmon, R and Mueller, S (1989). Allelopathic potential of sorghum-sudangrass hybrid. *Journal of Chemical Ecology* 15: 1855–1865.

Wondimagegnehu, M and Megh, S (1987). Allelopathic effect of (*Parthenium hysterophorus* L.) extract and residue on some agronomic crops and weeds. *Journal of Chemical Ecology* 13(7): 1739-1747.

Wu, H, Pratley, J, Lemerle, D, Haig, T and Verbeek, B (1998). Differential allelopathic potential among wheat accessions to annual ryegrass. pp. 567-571 in Proceeding 9th Australian Agronomy Conference, Wagga Wagga, NSW, Australia.

Wu, H, Pratley, J, Lemerle, D, Haig, T and Verbeek, B (2000). Laboratory screening for allelopathic potential of wheat (*Triticum aestivum*) accessions against annual ryegrass (*Loium rigidum*). *Australian Journal Agriculture Resource* 51:259-266.

Wu, H, An, M, Pratley, J and Haig, T (1999). Influence of temperature on the allelopathic potential of *Vulpia* residues. 12th Australian Weeds Conferences. Tasmanian Weed Society, Tasmanian.

Yaduraju, NT, Sushilkumar and Prasad, Babu MBB (2005). *Parthenium hysterophorus* – distribution, problems and management strategies in India. In In: Prasad T.V.R, HV Nanjappa and R Devendra (eds), *Second international conference on parthenium management*, University of Agricultural Sciences, Bangalore, India, pp. 6–10.

Zahran MA, Willis AJ, 2009. The Vegetation of Egypt, 2nd edition. Springer, Egypt.

Zhao, HL, Wang, Q, Ruan, X, Pan CD, and Jiang DA (2010). Phenolics and plant allelopathy. *Molecules* 15(12): 8933-8952.

Ziska, LH and Goins, EW (2006). Elevated carbon dioxide and weed populations in glyphosate treated soybean. *Crop Sciences* 46: 1354-1359.

INDEX

#

A

B

C

D

N

O

P

Q

R

S

T

U

V

W

X

Y

Z